T0200762

The ComSoc Guide
to Managing
Telecommunications
Projects

A volume in the IEEE Communications Society series:

The ComSoc Guides to Communications Technologies

Nim K. Cheung, *Series Editor*
Thomas Banwell, *Associate Editor*
Richard Lau, *Associate Editor*

Volumes in the series:

Next Generation Optical Transport: SDH/SONET/OTN
Huub van Helvoort

Managing Telecommunications Projects
Celia Desmond

WiMAX Technology and Network Evolution
Edited by Kamran Etemad and Ming-Yee Lai

The ComSoc Guide to Managing Telecommunications Projects

CELIA DESMOND

**IEEE
COMMUNICATIONS
SOCIETY**

The ComSoc Guides to Communications Technologies
Nim K. Cheung, *Series Editor*
Thomas Banwell, *Associate Series Editor*
Richard Lau, *Associate Series Editor*

IEEE PRESS

A JOHN WILEY & SONS, INC., PUBLICATION

For general information on our other products and services please contact our Customer Care Department within the United States at (800) 762-2974, outside the United States at (317) 572-3993 or fax (317) 572-4002.

Wiley also publishes its books in a variety of electronic formats. Some content that appears in print, however, may not be available in electronic formats. For more information about Wiley products, visit our web site at www.wiley.com.

Library of Congress Cataloging-in-Publication Data is available.

ISBN 978-0-470-28475-9

Printed in the United States of America.

10 9 8 7 6 5 4 3 2 1

CONTENTS

PREFACE

Telecommunications is a well-known, well-established and highly regarded industry that has been in place for more than 100 years. Over this time, the industry has been always growing, always evolving, and always playing an important role in the evolution of society in general. In most countries of the world, telecommunications has been deemed important enough that companies in the business have been encouraged by their governments to provide high-quality and reasonably ubiquitous service. Due to the high cost of equipment and associated work involved, governments put regulations in place to ensure that quality service is provided at reasonable rates for businesses and consumers who form huge customer bases for these companies. These service providers have generally been large companies that operate with some of the most sophisticated technologies and processes in the world. Of course, this requires high-level business planning and implementation. It also requires that tens of thousands of projects be undertaken each year worldwide, to keep the industry viable with new technologies, new service offerings, new mechanisms for operation, and new approaches. Projects have always driven the evolution of the telecommunications industry, which, today, is maybe most appropriately called the "electronic communications" industry. In this book, these names are used interchangeably to refer to the industry that provides voice, data, and multimedia services to both business and residential customers. Projects will continue to be very crucial to the success of the companies offering products and services in this area going forward.

The electronic communications industry is experiencing such rapid and disruptive change that in the early 2000s it is probably the industry with the highest overall rate of change. Within this industry, many players exist: very large established companies and small start-ups with innovative and clever products and ideas. Chapter 1 provides an overview of the evolution of the telecom industry over the later 1900s and the early 2000s, to illustrate the environment in which these projects are undertaken. This chapter also addresses some of the types of projects encountered in telecommunications. Chapter 2 discusses the importance of using project management within this environment, and Chapter 3 gives a general idea of what is needed for good project management. Of course, projects and the need for project management are discussed throughout the book. Next, Chapter 4 looks at what is needed to get a project going, looking at the source of the proposals and what might be considered in making the decision to go ahead. Stakeholders of these projects and how they look to the different types of companies in the business are discussed in Chapter 5.

For any of these companies to succeed they need to have a solid business proposition, strong products that meet the needs of the market, and operational capacity and capabilities to allow them to provide their services. The purpose of this book is to address the management of telecom or electronic communications projects. Although these are quite varied, and often have a very wide scope, the basics of project management can be applied to improve the chances and degree of success. These projects all occur within an industry evolving so rapidly at this time that it is almost impossible for any one person to even keep up with all developments. Implementing any project in this environment is difficult, but very exciting. And it is critical that project management concepts and techniques be applied to minimize the potential problems, make the work flow much more easily, and maximize the chances of success.

In the telecom industry, a high percentage of project managers (PM) are selected from the engineering ranks. The project teams, however, include people from many different departments and disciplines. Chapter 2 discusses the telecommunications industry environment and shows that the range of project types undertaken in this environment is extremely broad. It is not really possible to pinpoint a type of project, or even a set of project types, to label as telecom projects. Most telecom projects are technology based in some way, but many are related to the processes and the business environments in which the technologies are managed. Most involve working with or designing networks, services, or technologies; but to do this, people must have the corporate support and infrastructure enabling development of salable products and services as expeditiously as possible.

Some telecom projects are not strictly technology based. Even in highly technical projects, the team will need to understand the need for the service or product that it is developing or designing, the uses or applications of the product and the impact of the project implementation on its own company. The PM relies on the skills and knowledge of all team members, so project aspects will be considered from various perspectives. When mathematical concepts are presented, an understanding of mathematics equivalent to that attained by engineers is assumed. Many concepts are presented in a manner that allows easier understanding and use by engineers.

Why engineers? First, as mentioned previously, in telecom, as in other high technology industries, it is quite usual for project managers to be selected from the engineering ranks because most of technologies used are very complex and a solid technical background is needed to understand how they work, what they can potentially do, and the impact of many technology-related decisions. One of the main responsibilities of the project manager is making solid decisions for the company in all aspects of the project, especially technical issues or those that impact technologies. So it is critical that the project manager have this solid technical background. Next to engineers, probably the most likely choice for a PM is someone from an operations group, which usually also have a technical area. In telecom, operations tend to include all the background support needed to keep networks and services running, including trouble handling, testing provisioning, and other systems such as ordering or billing. Thus, people working in these groups generally also have a technical education, often an engineering degree. Even if the PM is not an engineer, engineers participate in almost every telecom project team. It is important for team members to understand project management as well as help reduce the frustration of overhead work, and just to understand the extent and value of work necessary to ensure that the project produces the best possible output in the required time, within the allocated budget.

Subsequent chapters look at other aspects of project management and how these can be handled in a telecommunications environment. Chapter 6 addresses project objectives and how these differ from business objectives. Chapter 7 talks about defining the project scope clearly and Chapter 8 outlines the procurement cycle. Chapter 9 reviews various aspects of project risks and Chapter 10 looks at different aspects of the planning for the project communications and the communications requirements. Chapters 11, 12, and 13 address some of the core concepts of project management: building and managing a schedule, the budget, and activities. Various aspects of handling people and the problems people face are addressed in Chapter 14. Finally, Chapter 15 rethinks why project management is important.

But, before any discussions become too involved, the scene must be set with some definitions to enable the reader to understand just what is being discussed. The first question that should be answered is "What is a project?" The answer to that question is found in a book called *The Project Managers Guide to the Book of Knowledge,* or *PMBOK,* published by Project Management Institute (PMI). The PMBOK defines a project as "a temporary endeavor undertaken to produce a unique product or service, with limited resources,

- Start and end dates
- Clearly defined objective
- Budget and other resource constraints
- Temporary team
- Perhaps initially defined deliverables
- Performed by one or more people, or organizational units."

So a project is not something people work on in their daily job but rather something that is undertaken for a defined period of time, to produce a defined outcome (product or service), which is different from anything existing. Generally, teams are multidisciplinary, although this is not necessarily the case. Certainly, it is more often true for project teams than it tends to be for day-to-day work, which is usually more focused within a functional department. Team members are selected based on the skills needed to complete the project. The outcome needs to be available, in full working order, by a date that is either predefined or defined before the project proceeds too far, and the budget also usually suffers from serious constraints. So we have a group of people who think differently, working together for a temporary period, to produce something different from their usual output, under time pressure. What fun! In order to do this effectively, project management is needed, and a defined discipline constitutes project management. The next question then, is "What is project management?" The answer to this question is also in the *PMBOK:* "Project Management is the application of knowledge, skills, tools, and techniques to project activities in order to meet stakeholder needs and expectations." This book discusses the knowledge needed, and many of the techniques or tools that can be used.

Just as food for thought, let us consider some initial examples of telecom projects. As we will see, the nature of these projects covers a huge range, but this short list gives some idea of the types of projects that might be encountered:

1. Design, install, and configure a network to support certain services.
2. Provide conversion plans to convert the entire network of an incumbent telecom company (telco), including replacing the technology and changing the architecture from circuit-switched to packet-switched.
3. Construct a new facility, data center, or POP.
4. Develop a new feature, product, or service according to a specific client's requirements.
5. Lay in a new fiber optic link.
6. Develop a new technology to enable the provision of new services.
7. Design a content-based, peer-to-peer application to run on the current high-speed Internet network.
8. Develop a new business model, including a new rate structure that encourages customers to subscribe to your company's services while at the same time ensuring that profit margin is maintained.

Clearly, the requirements for one of these projects differ drastically from those of another, and the skills needed to complete these projects also differ across the projects. Each of these is highly interesting, and obviously critical to the corporate success, so each needs to be managed in a way that will best ensure success.

CELIA DESMOND

Mississauga, Ontario, Canada
March 2010

ACRONYMS

ACWP	Actual Cost of Work Performed
ANSI	American National Standards Institute
BCWS	Budgeted Cost of Work Scheduled
CPI	Cost Performance Index
CV	Cost Variance
EAC	Estimate at Completion
ETC	Estimate to Complete
EV	Earned Value
FF	Finish–Finish
FS	Finish–Start
GPS	Global Positioning System
ICC	International Conference on Communications
ICT	Information and Communication Technologies
IEEE	The Institute of Electrical and Electronics Engineers
IP	Internet Protocol
IT	Information Technology
LAN	Local Area Network
NPV	Net Present Value
P & L	Profit and Loss
PDM	Precedence Diagram Method
PM	Project Manager
PMBOK	*A Guide to the Project Management Body of Knowledge*
PMI	Project Management Institute
PMP	Project Management Professional
POP	Post Office Protocol
PV	Planned Value
R & D	Research and Development
RBOC	Regional Bell Operating Companies
RFI	Request for Information
RFP	Request for Proposal
RFQ	Request for Quote

SF	Start–Finish
SPI	Schedule Performance Index
SS	Start–Start
SV	Schedule Variance
TDM	Time-Division Multiplexing
VOIP	Voice Over Internet Protocol
WBS	Work Breakdown Structure
WCET	Wireless Communications Engineering Technologies
WCP	Wireless Communications Professional
WEBOK	*Wireless Engineering Body of Knowledge*

CHAPTER 1

EVOLUTION OF THE TELECOMMUNICATIONS INDUSTRY

MONOPOLY STATUS

In order to gain insight into the telecommunications industry, first it is best to look at its history. Each country has its own way of handling this type of service, but there are many similarities in the treatment of telecom around the world. This chapter considers the trends of the evolution of telecom, which are mainly common around the world. Most of the background for this section comes from North America, and even in North America there were always some differences between the industry in the United States and that in Canada. The intent is not to give totally accurate specifics, since these vary country by country and also over time, but to show the trends that are essentially common to the telecom industry in most areas.

Initially, telecommunications service was treated as a "natural monopoly" essentially everywhere. The governments believed that this service, although very expensive to provide, was a basic right, and that it should be available to everyone at a reasonable price. Thus, in many countries, certainly in North America and in much of Europe, the government played a large part in the offering of telecommunications. In many countries, this service was provided by governments themselves, in a manner similar to the provision of other utilities such as electrical power. In North America, this was not the case, but the service was heavily regulated and many requirements and restrictions were placed on the companies that provided telecommunications. Regulation protected both the companies providing the service and the customers. Rates for basic service were kept low so that the service would be affordable by the masses, and requirements were often stringent, governing both the type and level of service that could be offered, as well as the rates that could be

ComSoc Guide to Managing Telecommunication Projects. By Celia Desmond
Copyright © 2010 the Institute of Electrical and Electronics Engineers, Inc.

charged for the service. Basic service was essentially voice service, and in North America the requirement was that local voice service capability should be offered at rates that were affordable to almost everyone.

In addition, companies providing basic telecommunications services were required to provide the capability of using the phone for immediate emergency support (medical, fire, and ambulance). This placed a requirement on the network that it be always available, so that someone who needed to place an emergency call would not be faced with an outage at a critical time. Since equipment and systems cannot be 100% reliable, the requirement placed on the telcos was that the lines and network be available 99.999% of the time. Thus, in North America the incumbent companies today provide very high quality telecommunications services at very reasonable rates, and almost everyone in these countries has access to telecom service. Because of this, most people have landline connections. This is not true in much of the developing world, where wireless services have blossomed due to the lack of affordable landline service in many areas.

Services that are not required for basic communication, such as data services, became known, generally through regulation, as enhanced services, or something a step above the basic service. These services, since they were not deemed essential by the regulators, did not have to meet the stringent criteria applied to basic communication services. So anything other than the basic landline voice services developed along somewhat different paths in many cases. But in many cases, these services were actually offered by the same carriers who provided the basic voice service, under separate regulatory rules.

Local voice service was provided to allow callers to place calls to others within a local calling area, generally the municipality in which they were located. In addition, these local networks were interconnected to each other, allowing callers to place long distance calls to people in other municipalities and even other countries. The interconnecting networks are often called backbone networks. In the United States, there were companies offering local service within towns and cities, and also interconnection between these towns and cities as long distance service. The names and sometimes the nature of these companies changed over time. Perhaps the largest ones were most notably known as the Regional Bell Operating Companies, or RBOCs. These companies were in turn interconnected with each other by one large interconnector called AT&T. The local calling capability was considered to be basic, whereas the long distance calling was discretionary. So the providers were allowed to charge more for the long distance calls than for local. Given the stringent requirements on the networks and the rates for local calls, companies were generally providing local service at a loss. They could balance this loss by making money on the long distance calls. However, even here they were restricted by the regulators. The provision of data services was also not basic, but data service was also regulated, and in the early days, not a large enough part of the service providers' business to really matter.

Because of the rules and restrictions, and also due to the high expense of providing such service, telecommunications grew to be a monopoly service almost everywhere. Even when there were other providers on the scene, they were generally either limited in what they could provide, or in the volume that they were allowed to provide, protecting the interests of the monopoly provider.

Thus, initially, incumbent providers were heavily regulated and provided basic local voice services at low rates. They generally also provided long distance calling, usually at rates that were higher than the costs, to cover for the noncompensatory local services. And many of these companies also provided some other services, generally in small amounts, to make additional profit. This was the trend almost everywhere for many years, even though the specific rates and services offered varied over time and location, and by company. As mentioned earlier, trends were generally similar around most of the world and in most companies, although the changes that occurred happened at different times in different locations. Initially, also, the incumbent companies were government owned, or at least government controlled, which caused some differences by country. Most incumbents provided mainly voice services and small amounts of data. Generally, in the late 1900s the networks carried about 80% voice and 20% data services. And until the 1980s, most of the voice services and some of the data services were carried on analogue networks, in both the access and the backbone. Data services were a mix of analogue and digital until the late 1980s when the trend was to convert networks to digital technologies end to end.

COMPETITION FOR LONG DISTANCE SERVICES

In North America, new technologies that allowed different methods of carrying traffic and also changes in the attitudes of those using the services caused pressures to begin attacking the business models. Initial attacks targeted the long distance services, since the incumbent telcos were making good profits on these services. Others understood the economics of these services and wanted part of the action. They recognized that they could provide this service at low cost and charge rates much lower than those charged by the incumbents, still making a profit. So the initial successful attacks on the established business of the incumbent telcos targeted lucrative long distance services. These attacks came from new long distance providers, resellers, and value-added carriers who leased network capacity from the incumbent telcos and used it to provide services of their own. Today, everyone is quite familiar with this evolution and the competition has driven the rates for long distance to a fraction of what they were 10 or 15 years ago. As we will see below, other factors are also impacting these rates, as everyone is also aware.

The next stage in the trends was the change in the providing of data services. Initially, except for the earlier telegraph services, data was carried, by the incumbents at least, on the same analogue backbone as the voice services. Data entered the access network via a modem, which had an analogue output, or a dataset, which outputs signals in a digital format. From there, it traversed usually private lines, sometimes multipoint lines allowing many terminals to share a port on a host, or it operated in a switched mode along the same (circuit switched) lines as the voice services. In the 1970s to 1980s, this started to change with the advent of packet switching.

The access and the backbone of the telephone network were designed to accommodate the patterns and needs of voice traffic. The network is optimized as a network that receives a call on a single circuit, checks the destination number for that call, and sets up an end-to-end circuit for the duration of the call, dropping this at each step along the path when the call has been completed. This type of network is called circuit switched, and huge numbers of man years of research and implementation went into optimizing the circuit switched networks to very efficiently serve voice traffic. Much of this development was done by Bell Laboratories in the United States, where hundreds of research projects were underway at any time.

Around the 1970s, packet switching technology started to emerge. This had been used earlier for private and small networks, and it was optimized for data traffic. This traffic did not need to have a circuit established end to end for the duration of the call and, in fact, given the nature of the traffic, establishing an end-to-end circuit was quite wasteful. Packet switching technology began to be deployed widely around the world by data services providers and also by incumbent telcos. Initially, these packet switched networks served only specific data services—digital data services—and the traffic volumes were low in relation to voice traffic volumes. Some data continued to be carried on the old analogue networks, with certain services being carried on the packet switched networks. During this period, the Internet started its well-known growth and, of course, that growth continues even today. As the Internet expanded from researchers and universities initially, then to companies, and finally to individuals, it was to be expected that the amount of data traffic and the routing patterns of that traffic would change drastically. And, at least initially, this Internet traffic was data traffic. So by the year 2000, that 80/20 ratio of voice-to-data traffic on the network was changing to a much higher percentage of the traffic being data. In parallel, technology development enabled video and multimedia traffic to be carried on the network. And as these enhanced services spread, the amount of traffic and especially the amount of nonvoice traffic continued to increase.

COMPETITION FOR LOCAL SERVICES

One of the next developments was the attack on local service provision by many other companies. In some cases, this was cellular service, since mobile companies had been growing even in North America, where essentially everyone had access to reasonably priced land line service. By the mid 1990s regulation allowed companies other than the incumbent telcos to offer local service, so other companies started to move into territory previously covered by telcos. New companies sprang up and other utilities, such as cable companies and even power companies, also decided to offer data and voice services. At the same time, telcos requested permission to expand their service offerings to include those offered in the past by other providers. Specifically, telcos requested, and often gained, permission to provide broadcast video service.

The net result of all of this was the creation of competition in every stronghold of the incumbent telcos. By the time this happened, they had already been struggling to change their internal culture and focus to ones equipped to deal with competition. The competition for local service and for the provision of data services just heightened that requirement.

Initially, companies offering competing local services positioned these as secondary services to avoid the requirement that they provide emergency call service. Over time, developments in the mobile service technologies have made it possible for some of the emergency capabilities to be provided even from mobile phones. Specifically, it is now possible to locate someone with a medical emergency within very short distances, even if they call for help using a mobile phone. This ability is one of the reasons that many people today are quite comfortable having a mobile phone as their only phone, even in areas with the well-established, high-performance, and low-rate landline service

As this evolved, the nature of the traffic carried on the networks also changed over time. As all facilities became digital, all traffic, whether it was initially voice, data, or video, was carried in a digital format. While it is still possible to determine the nature of the traffic at the end points (and, with the right protocols, information identifying the traffic type can also be carried through the network), once traffic is converted to bits, it becomes less relevant to distinguish the difference between voice and data. And when we add to the picture other traffic such as video and multimedia, the picture becomes even more cloudy. Even though the actual nature of the traffic might be known, the fact is that as it travels through the network it looks like data, so the most effective mechanism for carrying this traffic is packet-switched networks.

COMPETITION STARTS TO SPREAD

By the early 2000s, telcos faced serious competition to all aspects of their business from many others, most of whom had not been on the horizon of telecom a decade earlier. These other companies were starting into the telecom business from different starting points. Some had established networks, as the telcos did, which reached most or at least many of the clients interested in voice and data services. However, most of these established networks had been initially designed for other services and were not optimal for the provision of two-way, high-quality voice and data services. So those other providers with established networks faced the problem of high costs of conversion of their networks to ones that could effectively provide the voice services.

INTERNET AND MULTIMEDIA DISRUPT THE BASIC NETWORKS

By this time, the Internet had become almost a commodity in itself and multimedia services were blossoming. One very significant development then was the use of the Internet protocol (IP) to carry not only the data services, for which it had been optimized, but also voice service. Why not? The voice traffic in the network was already in a digital format. Of course, the transmission requirements for voice differ from those for data, as they do for video and other forms of multimedia. So the introduction of Voice Over IP (VOIP) was not a simple, straightforward transition. But, for someone who had no established high-cost, high-performance circuit-switched network, it made more sense to introduce packet-switched networks, which are better for carrying the new forms of digital traffic. And by this time, the network traffic patterns had evolved to a point where the traffic was no longer 80% voice. In fact, depending on how one defines traffic, the voice traffic will soon be (or maybe already is) 80% other than voice.

In parallel, new and very different providers appeared on the scene, developing vastly different, but also wildly popular, new services which could not even have been imagined a few years earlier. Internet services had blossomed through the 1990s, and they continued to grow even through the demise of the hundreds of .com companies that initially sprang up to take advantage of the predictions of huge market share by providing data and multimedia services. Music was one of the initial multimedia services, and some of the initial service providers tested the regulatory and legal environments in which these services could be provided. Legal issues also abounded related to some of the less savory Internet services, and these still continue today. Some such services were actually illegal in some countries, whereas others just raised questions about what should and could be allowed. Probably, many people remember the Napster service, which provided a platform for end users to share their files. In the case of Napster, the files shared were usually music files. What this amounted to, essentially, was one end user downloading a music file from somewhere. The initial file was probably paid for by the user, so the musician and the user were both happy. But then Napster provided a platform for one user to check out the music owned by another user, and people started sharing these files amongst themselves for free, cutting the original artist out of revenue for his or her product. As expected, this did raise legal issues and, in the end, Napster had to stop this service. However, the next generations of this type of sharing capability were not such easy targets for the legal systems. Kazaa, a peer-to-peer-based application, allowed one user to look at the contents of another user's computer, and they shared directly, without middleman control. Whereas it is certainly possible for legal systems to determine who is sharing the files that are IP protected, and they might be able to win in court, the sharing by individuals is generally too small to warrant the cost of court action, and to complicate things further, there is no middleman to sue as there had been with Napster. It can certainly be possible to put some sharers out of business, but in the time it takes to prosecute one or a few transgressors, hundreds more spring up. So the landscape has changed from a service perspective as well as in other ways that make service decisions more complex for the service providers. Think about the issues that face project managers charged with implementing projects with such services included. Not only do they face the usual project and service dilemmas, but they also have to spend time and energy worrying about how best to manage potential legal and ethical issues.

Electronic commerce evolved throughout this time and continues to do so. Not far behind came many other creative offerings, including Skype, Second Life, and Facebook, to name just a few. Skype was one of the first big VOIP services, offering voice calls totally free to anywhere in the world as long as people were willing to talk through their computers, or for tiny fractions of the rates charged by the telcos (even after the huge rate reductions mentioned above) for calls terminating on actual telephones. Second Life is essentially a real-life video game in which those who wish to set up a personality, called an avatar, can interact in real time with other individuals, and, more importantly, with the companies and others who purchase space to offer their electronic (and sometimes real) wares over the Internet. Many companies have already established their presence in this virtual world. Facebook, which followed an earlier version called MySpace, started initially as a mechanism to allow college students to share rich communications with their friends. (MySpace was initially aimed at high school students). Facebook expanded their market to include anyone, and it is a good example of the new types of services. Although the Facebook platform was developed with a number of features, many developers who have nothing to do with the Facebook developers have added additional applications that Facebook users can use to enrich their experience on the site. The users pay nothing to use the facility; it is supported by advertisers who post ads on the site. Generally, they pay by the number of clicks, so most of the applications are designed to encourage a lot of activity on the site. Many cute, fun games are featured, which the users share with their friends. Of course, all of these applications generate a lot of network traffic, and most of this traffic is data. Even the voice traffic is carried as VOIP, which is best carried by packet-switched networks. Thus, there are not only new services to think about, but the services are built using a new model, which is much more open than that of any of the services that were offered over the previous history of telecom. And at the same time, the service providers must consider whether they want to offer their services based on the long-established models of either pay per use or pay a flat rate for some amount of service, or whether they wish to move to a new business model in which the revenue does not come from the user at all.

A NEW TELECOM ENVIRONMENT

The implications of this traffic evolution have become evident—the established network provided by the large incumbents is no longer optimal for carrying the bulk of the traffic. Thus, the industry has evolved to a position in which the telco incumbents face many problems. Every service they have offered for years as a monopoly is no longer protected, each attacked by competitors. Competition is appearing from every direction at the same time, mainly from sources the incumbents have difficulty in predicting or sometimes even identifying as competition. At the same time, the billions of dollars that they have invested in networks, both to carry the actual traffic and as back-office systems to support the business, is becoming less and less relevant to providing current services. Incumbent telcos face problems that are probably bigger than those faced by other utilities moving into the provision of communications services because they have to essentially remove all the established networks, replacing them with packet-switched networks. This involves many extremely large technology-based projects and also many more that involve the creation of new processes, procedures, and business strategies. Not only must the networks be replaced, but also the provisioning, ordering, billing, and other service-recording systems must all be converted as well. These would be interesting projects under any conditions, but they must be carried out at an accelerated pace, at the lowest possible cost, producing extremely high-quality products, in order to maintain their customer bases in the face of extreme competition.

These telco network conversions are much bigger projects in terms of time, cost, and impact than these companies ever faced before. In parallel with all the primary business conversions, most of these companies also need to address their internal corporate culture, since the way in which they have conducted business for many years will not generate success in the current environment. Thus, projects often have components of all the standard business areas such as engineering, operations, IT, marketing, and sales, as well as additional components addressing the business attitudes and methodologies.

So, what type of projects might the service provider undertake? Let us consider but a few of these:

- Developing a new service
- Developing new features for an existing service
- Analyzing the introduction of another company's competing new high-speed access service, enabling the service provider to determine the best competitive response
- Work with a major national customer to implement his network in a way that gives him significant savings, while at the same time improving his service by moving him onto a new broadband network with better management capabilities.
- Design, implement, and manage a network within a conference complex for a group of UN leaders who will be attending a meeting in your city. The communications includes incoming and outgoing voice and data calls to and from the complex, internal communications amongst the politicians and their sup-

port staff while in the complex, plus a secure LAN within the meeting room itself, which allows each politician to communicate and share files with his or her own "sherpa" during the meeting.

- Managing a serious cable cut in a remote area, in which the cable was carrying over 40% of the national backbone traffic and the redundant backup facility is currently being upgraded and, therefore, cannot reliably carry its full traffic capacity
- Implement a new IPv6 capability in a separate network for customers willing to move to the leading edge protocol.
- Equip the current network with a new billing system that is more flexible that the current one, allowing new rating models to be adopted when desired.
- Training all current employees on the new network model and its benefits
- Move all customer service for medium business clients to one common national call center
- Introduce a new culture to the employees that is more conducive to determining customer requirements clearly before initiating design of a new service and provides them with the tools to be able to do this.
- Learning to be more effective at marketing product and service lines

Thus, even within service providers, the variety of projects, and their requirements for skills, knowledge, and people, the inputs and deliverables are extremely varied. This makes life very interesting for project managers and team members in these companies.

So far, we have mentioned only telco service providers and research facilities such as Bell Labs. While this part of the business is obviously huge, these companies in no way comprise the full telecom industry. There exists a complete continuum from the people who create the initial ideas for new technologies and designs, through the people who design and build technologies, networks, and support systems, to the companies that provide direct end-user communication services, whether voice, data, multimedia, or value added, and also to the governments who set regulations and policies, and control things like spectrum allocation.

So at one end of this continuum we might find companies that do basic research. Bell Labs was an example of such a company over the years, although today not many companies are funding the basic research that enabled the telecom environment to develop many of the highly innovative technologies and designs that made the industry so successful. Also, somewhere in this continuum we would also find companies that do the research into potential new technologies and products, and who also provide assistance in the technical and management aspects of integrating these upcoming products into the current networks and environments. We can include these under the equipment vendor heading, although not all such companies fall into that place in the overall model. These companies face the issues of setting the standards for the technologies, often in parallel with their initial product development. So their world is changing as described above, and also changing due to their own work in the standards areas.

Of course, there are also the long-time incumbent telcos in the line, as we have discussed above. But in the service provider position, we now also have many, many others who fit very different profiles. There are the new entrants to the network and/or service business. These companies may already have networks, such as cable companies or power companies. Alternatively, they might be companies building new networks, which would naturally use the most modern technologies and designs, optimized to carry the current and future services. In addition, there are the new entrants to the value added business such as those who develop new applications and features for Facebook services.

Corporate cultures differ from one company to another. The business goals are different, as well as the equipment they own and services that they provide. Thus, the projects will be different. Even if two companies develop similar services, they do this under different conditions, within different cultures, using different technologies and with very different business goals. All of this has to be kept in mind when viewing any telecom project.

One thing, though, that is evident is that there are many, many exciting and critical projects underway and on the horizon for any company participating in the electronic communications business.

WHAT ABOUT THE FUTURE?

As we move forward, telecommunications is becoming less an independent infrastructure and more a part of the overall ecosystem. Telecommunications capabilities are starting to become an integral part of activities in almost all fields of life. In medicine, many governments have huge projects underway, not only to electronically store and work with data, but to use communications equipment as one of the tools in performing operations and other activities. In many other areas of life, sensors and monitors will allow humanity to track, record, and even modify conditions that endanger the future of the world. This is true in the case of water use in agriculture, reduction of pollution, and control of automobile traffic. The projects that will design and implement these technologies will be extremely exciting, with huge potential gains for individuals, corporations, and society. We see huge changes in entertainment and commerce all enabled by electronic communications. The scope of the projects will be even broader than they are now, and the stakes for success will be even higher. With telecommunications infrastructure playing an enabling role in world economic development, the projects will be highly visible to other disciplines and, in many cases, around the world.

All of this means that projects are here to stay for the foreseeable future, and success is critical. Project success is much more likely when proper project management is used. Thus, project management is needed.

CHAPTER 2

WHY IS PM IMPORTANT, ESPECIALLY IN TELECOMMUNICATIONS?

Why is it necessary and even important to use project management? People who appreciate and practice project management will say that it is always important to use project management tools and techniques for projects of any level of complexity. One might then ask why they would believe this. There are many reasons.

Some of the benefits of the application of project management techniques that are important for any project include: ensuring good teamwork and other people factors, meeting the project budget and schedule, producing products and work of the expected quality, effectively managing project risks, ensuring that good communication occurs, and ending the project by delivering results that include the full scope as originally planned. All of these things are aspects of management of a project and they are the factors that define success in any project undertaking.

For any project, there are many factors in play. First, consider the project team, comprising the group of people actively working to deliver the product or service that the project is in place to produce. The project team is generally a multidisciplinary group; the people on the team will have different backgrounds, different objectives, and different ways of thinking. To ensure that such a diverse group can produce whatever is required will take some management focus, and that focus is one of the most important aspects of project management.

TEAM DIVERSITY

The team will also usually be temporary, composed of people who do not usually work together. Thus, it is very likely that they do not know each other well, creating a need for time and effort to be used to develop understanding of other team members, to ensure that people will understand what is needed to work with the other members of the team. Given the multidisciplinary aspects of the project team, we can see that considerable management effort will be needed to understand the diverse points of view of the people assigned to the project, building a team that will work together well and with effective communication to achieve the project goals. Again, this skill is part of project management.

In any project, the people on the team need to evaluate their own environment, their own project, their own skills, and their own company. Based on factors such as these, they can decide how they can best work with the rest of the team to produce the best quality product or service possible.

RESOURCE LIMITATIONS

Another characteristic of projects is that there are usually hard limits to the budget and other resources required to implement the project. Thus, there will be a need to prioritize just how and on which components the money should be spent, which is again a component of project management.

TIME CONSTRAINTS AND LIMITATIONS

Time is always an important factor for the project manager. Since projects take people away from their regular jobs, there will be time pressure to return the people to their own organizations, even when there is little time pressure to make the project's product or service available by a specific time. However, most projects start with schedule constraints already in place, so time management, another key component of project management, is also a crucial skill. Any team will be better equipped to meet both time and cost targets if these are clear, reasonable, understood, and well communicated.

RISK MANAGEMENT

The understanding and management of risks is crucial to the project manager. Considering all of the above factors of the project environment—people working with others they do not know on multidisciplinary teams to produce something that is initially not well defined and outside the normal work framework, with time and money pressure—would it be reasonable to expect that the environment for any project is not risky? Obviously, skill in risk management is always needed.

ENSURING QUALITY

What about quality? Projects are undertaken to take advantage of business and technology opportunities to introduce a new product or service. In most cases, a new product or service displaces something else that had previously met the needs of the targeted market. The quality of the new product or service is often the factor that will allow it to compete well with the previously established competing product or service. To make the new product or service of the best possible quality, it should be produced using quality-management techniques. Quality management, therefore, is another important aspect of project management.

SCOPE DEFINITION

One of the most important aspects of project management is the understanding, clear definition, and effective management of the project scope. Consider the nature of a project, which is to produce something unique. The output of a project is often not something that people already know and understand. In the case of ongoing work in a normal production environment, it could be quite reasonable to expect that people understand what needs to be done and what is to be produced. But at the outset of a project, it is very likely that no one except possibly the project sponsor and the project manager has a very good understanding of what is to be produced, let alone how to achieve this. In most cases, quite a lot of work is required to clearly define the required output and the work needed to allow this to be provided. Scope definition and subsequent management is a large part of project management.

One of the Project Management Institute project management process areas is scope management, and within this area there are processes and tools that are used to clearly define the scope of the project and to manage proposed changes. It takes just common sense to realize that any project team can avoid mistakes by ensuring that all those involved with the project, either as workers, managers, or receivers of the product, have a common understanding of what it is that the project will deliver, and it would be even better if they all had the same understanding of how the production of this end product would be done.

Once the project scope has been defined and approved, and the rest of the planning has been done, project implementation begins. At that point, the scope, the team, the schedule, the budget, and other project parameters have been set. After this point, if changes are needed, there is potential for this change to impact many aspects of the project. Such proposals for change are called scope changes. There has never been a project that did not have changes to scope. Some huge projects have experienced literally thousands of changes to scope. For the most part, people do not propose scope changes for the sake of the change itself; there is generally good reason for wanting the change. And in some cases, it makes much more sense to implement the change than to not go ahead with it. But, making any change impacts the project, and too many of these can cause a project to fail, no matter how well everything else is done.

Scope change requests come in many forms. In one form, someone (usually the customer) says, "Just add this little feature. It only costs a little bit more, takes a little more time, and everything will be 100 times better." That might be true and it might well be much better to incorporate the small change now, while there are people geared up, equipment in place, and so on, than to go ahead without doing it and try to incorporate it later, outside of the project. It might be possible to do one, or two, or three such additions that will cost little in time and/or money, but when the number of these requests gets to ten or a hundred, there is no way within the time frame and budget, and with the people on the project team, that these can be accomplished.

Let us be ultraconservative and suppose that each proposed scope change was accepted and the time required to complete each one was only one person-hour. Most people will think, rightly, that expecting each request for change to take only one person-hour to implement is unrealistic, as scope changes often take weeks or months to accommodate. But for this example, suppose each one only took one hour. If we had 1000 requests for "tiny" changes, which is not really unexpected in, say, a project with a team of 50 people running for, say, 18 months, accepting all of these requests would mean adding 1000 person-hours to that project. That's an addition of 125 eight hour days to the project, or more than an additional four-person months. And if this project received three times this many requests, which is not all that unusual in some environments, we would need an additional person-year to complete just the things that were not in the initial plan if none of them took more than one hour to complete. But of course, most take far more than a short time to complete, and every project receives these requests, usually many of them, so there needs to be some time included somehow to do these unplanned items.

Therefore, in order to be able to meet the schedule and the budget, it is necessary to incorporate some mechanisms that will allow the project team to deal with the many change requests that will inevitably arrive. These need to be considered, assessed, and either accepted or rejected, and, regardless of the decision, the impacts of that decision must be dealt with. All of this will take time, and time costs money. So a change request process is a very important component of project management, and one which is needed for every project.

In another situation, the change request arrives via a statement from one of the team members that there is a design flaw in the project or the product to be produced, and unless this is fixed the whole project will be down the drain. It just will not work. This is quite different from someone thinking up a new addition that would be nice to have. This one, if the requestor is correct, is necessary if the project is to be successful. But it is still a change to the project scope, and making the change will cost time and money, not to mention the possibility of needed additional skills or a new risk to the team. The change must be accepted, but this should not be done lightly. Good project management practice requires that this proposed change go through the change request process, and if accepted, be implemented only once all the impacts are understood and the required resources have been obtained.

In short, every project will experience proposals for changes to the initially planned scope. And these changes do interfere with projects. A few smaller ones might be incorporated by having people work a little harder, or some such mechanism, but, overall, if you take on many changes, or large ones, the project will not finish on time, with the right quality level of product, at the right budget. Well-defined project management techniques help the project manager be clear on what factors the team, and the project manager himself, is being measured. Quite possibly, one's bonus for the year, or even keeping one's job, depend on projects finishing on time, on budget, or with some specific deliverables working well. If as a project manager you keep taking on more work without ensuring that compensating budget or schedule changes are put in place, failure is inevitable. Every project undergoes scope changes; we know before the project starts that these requests are going to come, so it only makes sense to plan for them.

PROJECT OBJECTIVES

Another aspect of project planning is the setting and communication of clear, attainable, and measurable objectives. Doing this can avoid frustration by ensuring that the team members and the key stakeholders all march to the same drummer. A team will be better able to ensure that all those involved with the project in any way have the same expectations and the same information if there is good communication, especially agreed-upon formal communication. But communication does not just happen: proper communication requires planning and focus, both of which are skills of the good project manager. With the right attention by all team members who have something useful to contribute, decisions can be made consciously and for the right reasons. It is part of the project manager's role to ensure that this will happen.

With the right communication, it is easier for the team to avoid known pitfalls, even for a fairly straightforward project. Consider the following example of organizing an IEEE conference.

In a volunteer organization such as IEEE, typical projects might be the organization of conferences, whether small ones of only 50–200 attendees, or large ones attracting thousands of people. Such projects include the preparation of publications with papers by many authors; the initial request for these papers; receipt and review of the papers; organization of the papers into sessions with a technical theme; the organization of meals and coffee breaks; the organization of events such as receptions; providing speaker instructions or awards presentations; the making of hotel arrangements, and, possibly, conference center arrangements so that attendees will have a place to stay and sessions can be held; the publicizing of the event; and so on. Obviously, these functions are not all carried out by only one or two people, so there is a requirement that there be good communication amongst the organizers. The rooms need to be the right size for the sessions, and the conference needs to have enough rooms. The attendees need to have the information about the conference, how to register, where it will be held, when and where to attend the sessions, and so on. So not only must there be good communication amongst the organizers in each area, there must be good communication amongst the organizers of the different aspects of the conference, and excellent communication with potential and confirmed conference attendees, to attract people and ensure that they get the most out of the conference. The people doing the organization are all volunteers. They work for different companies and frequently live in different countries, yet they must work together to make the conference happen smoothly. This can be done only with strong, planned, well-organized communication.

Another quite different volunteer project is our example of the Communications Society Wireless Communications Engineering Technologies Certification, for which the team has over 100 members. This project again has not only a large team, but people working in very diverse areas and roles within the project, and these people live in over 20 countries around the world. Although it is not necessary for each of these people to communicate directly with all of the others, it is clear that the certification cannot be developed efficiently or successfully without strong, planned, organized communication amongst the team members, and a great deal of it, for a successful outcome. In volunteer work, project teams are made up of people who work for companies, self-employed entrepreneurs, and academics, who frequently live in many different countries around the world. These people agree to do the volunteer work in addition to their own workload, for no monetary compensation. The project manager does not have any control over the people in a volunteer organization. These people are working on the project because they want to be there, and if they do not feel like providing their deliverables until the last minute or submitting a status report, they do not have to. And unless the project manager can influence the people positively, some of the material needed for the project will not meet requirements. Today, similar situations often exist in electronic communications projects within for-profit companies. The team members are not volunteers, but they may be working in environments that are as diverse as those described here and they could well also work for different companies that are cooperating to build a joint product or service.

Managing projects for volunteer work is different from doing so at work, where the project manager has a degree of defined authority. But even in work situations in which project managers have control over all the people on their teams, generally they do not have control at all over most of the project stakeholders. Project managers do not always directly supervise the people that are working on the project. In some management structures, the project manager is more of a coordinator, and does not directly supervise any of the team members. The team members continue to report to their usual supervisors, but are assigned project work instead of, or in addition to their regular job. The project manager is charged with overseeing the project work and making sure that it all happens according to the project needs and plan. That means that it is up to the project manager to use relevant skills, not the limited authority associated with the project manager title, to get people to do things, to get them done well, and to get them done on time.

WHAT ABOUT TELECOM PROJECTS?

The title of this chapter emphasizes the importance of project management for telecommunications projects. So far, the discussion has been about the importance of project management for projects in general, with no specific reference to telecommunications projects. This book focuses on the application of project management principles to telecommunications-related projects. It is more important to use proper project management technique in telecommunications than in most other sectors due to some particular characteristics of the industry.

Consider the type of projects that typically occur in the telecom industry. Many of these involve huge networks, either for a provider or a large end user, or extremely complex services and equipment, with hardware, software, business, and integration aspects. Teams tend to be large—from 25 to hundreds of people per project—and the technologies involved are extremely complex. Project management lends itself well to handling such size and complexity. Most significantly, in this rapidly changing industry, many companies find themselves needing to do things for the first time, and handling these situations by implementing projects is often the only practical way of accomplishing this.

Telecommunication technologies involved in projects are usually fairly new (and, therefore, not well known), and there is a requirement in many cases for interworking of many different technologies. All of this creates a requirement for significant technical knowledge on the team, and, generally, also significant business or marketing knowledge, making it necessary that the teams be multidisciplinary and competing for people with scarce and valuable skills and experience. The application and interworking of new technologies ensures that telecom projects will contain a higher than average degree of risk. This unpredictability implies that there will be challenges in defining the project scope accurately, in turn making timelines and budgets hard to nail down. All of this adds up to a need for strong and ingenious project management.

Telecom projects are planned and implemented in an environment that experiences continuous, significant, and rapid changes. The following is a view of several aspects of the telecommunications industry that are important to the application of project management in this sector.

Technologies

Whether we look at access, transmission, switching, terminal devices, service platforms, servers, billing, provisioning, ordering testing, or any aspect of telecommunications products or services, we find very many technologies in use. These are at various stages of development, and they all must interwork with each other and with applications that customers create, such as local area networks. Consider access technologies. These could include copper pairs, video cable, wireless, or fiber optics. And if we take just one of these technologies, say wireless, there are probably no less than 15 different versions of this technology that might come into play in some way in a project, including WiFi, WiMax, satellite, Bluetooth, and ultra wideband. Some projects do not involve more than one of these, while others might involve many. But almost all telecom projects do include at least one relatively new and rapidly evolving technology, resulting in an unstable project environment. Thus, there is a very great need for technical skills that must be developed and kept current as the project evolves, placing more stress on the project than there is in other industries in which the technologies are less complex and more static.

Services

In this book, we discuss the changing nature of the services offered today, as the Internet and entertainment media become integral to electronic communications services. Whereas for many years a telecom service involved mainly local voice service or long distance voice service, with possibly a separate data component, today's services need to be built by integrating voice, data, and multimedia in innovative ways in order to be successful. The requirements for such projects are much more complex than they have been in the past for telecom services, since the need for innovation creates higher risks, as does the integration of the many different components. Customer expectations of new services are escalating very rapidly. Capabilities that were considered close to miraculous five years ago are now considered outdated and of little interest. It is extremely difficult, despite the efforts of armies of marketers, to predict what service will be a hot item months in the future. So the risk is high, and the teams of people from many different areas must work together in order to introduce successful service offerings.

Companies in the Business

In the previous chapter, we talked about the rate at which competition is escalating in electronic communications and the many different companies that are now providing these services. Projects are put in place to either take advantage of opportunities or create solutions to problems. These opportunities and problems occur in an environment that is ever changing, where competing companies may create market or operational pressures that impact the project, the project work, and the project outputs. If the home company of a project merges or becomes acquired, the nature of the project is often affected. When a new competitor appears with products that are new, better, or different, priorities of the project change. So the evolution of the companies in the relevant business sector places a heavy strain on the project teams working to provide services and products for telecommunications.

Regulatory Environment

Over the past 10–20 years, the level of regulation governing telecom has been gradually decreasing in most areas. However, many things are still regulated, and some

are heavily controlled. Once regulations are set and understood, they place requirements on the project that must be incorporated into the project requirements along
with any other requirements. Project teams often need to work during intervals
when regulatory changes are expected or threatened, without knowing what the situation will be at some point in the future. This greatly increases the project risk, and
also the stress on the project team, creating a need to manage toward the most optimal solution.

Successful Business Model

Since the inception of the telecom technologies and business, the industry has operated under similar business models in most countries. These business models have
been based on the premise that the customer pays for the service delivered. Sometimes the models were usage based, and in other cases the payments were flat rates,
usually monthly. In some cases, the customer pays for both outgoing and incoming
calls, whereas in other cases only outgoing calls are charged. When there is a need
for equipment beyond that used for standard service, the customer might either
lease or purchase the additional equipment. Today, for many services these traditional models do not apply. Instead, the services are sometimes offered free, with
advertisers paying the cost, with rates depending on the popularity of the services.
Thus, in addition to the need for more innovation in the development of new services, the teams often also have to deal with a new business model. This again
makes the environment more risky, increasing the need for good project management.

Internal Corporate Structures

Yet one more issue that teams face during today's projects is the internal restructuring of their companies, moving from highly hierarchical structures to ones that are
flatter, or from structures based on older services such as local, long distance, and
cellular to structures based on models of newer services, such as video or social networking. The team then is essentially standing on a rolling platform while they
work on the project. This raises the stress level of the people on the team, including,
of course, the project manager. People will dilute their focus on the project as they
watch the developments in their company, increasing the pressure for the management of the project.

Customers

Given that competition is rampant today, customers have to deal with more choices
and the anxieties they bring with them. They do not know with any certainty which
company to buy from or which offer to take. If the project is providing a product or
service for outside customers, there is a need that is becoming more important for

telecom projects—that of understanding the perspective of the potential customers and even determining what emerging entity might become a potential customer. This is an increase in the scope of projects above that experienced in this industry in the past requiring new skills and teams to deal with new perspectives. This, in turn, creates more need for team building, for clear understanding of the project scope, and for communication of this changing information to the project team. All of these are again components of the project manager's job.

The Best Way to Market

Customer needs are changing, as are business models. Even the types of services and products that should be offered are not what they used to be. There is also a need to for the project team to understand how to build marketing plans, and how to best approach the market. This again extends the skills needed and the number of perspectives the team must consider.

Service Models

Telephone services have been highly controlled in the past, in the sense that the intelligence that makes the service work, and the control of the network, its traffic, and all of its operating parameters have been all completely in the hands of the providing company. Traditional companies are very protective of these services, and their control, but this way of doing things is becoming less dominant in the telecom industry. Newer services use open platforms, and often contain components provided and built by multiple providers. Network intelligence and control is no longer solely resident in the core of the network and, in fact, much of the control resides at the edge. Rather than a service being delivered in its entirety by a single service provider, it is now often delivered by a loose partnership of specialized companies. This, of course, increases the risk that the service will not work well, and the pressure on the architects of the network to know, understand, and work with multiple technologies and providers. Again, this is one more pressure on the project, and one more reason that management is needed.

Network Architecture

In the previous chapter, we discussed the evolution of telecom networks from circuit-switched systems connected via TDM facilities to packet-switched networks running Internet Protocol. This change of the technical environment in which new projects are implemented from that which many team members know drives a need for constant learning on the part of the team members to keep current. Inability to do so reduces the technical effectiveness of the team, adds risks due to the higher probability of wrong decisions arising from unfamiliarity with the technology, and also can seriously impair the self-confidence of the team members, leading to personal stress and highly risk-averse behavior. Identification and implementation of the necessary training should be a factor in the planning of the project.

CONCLUSION

It is clear that the skills, technical and otherwise, required to effectively complete a telecom project today are more varied than those that have been needed in the past. Summing up all of the variables discussed in this chapter and in the previous one, we see that many, many things are evolving and changing, and the sum of all these changes is a very volatile environment in which to do projects. Because of the degree of change encountered in the telecommunications industry, the need for project management is much greater than in many other sectors. Electronic communications project teams are operating with a high degree of change in many areas, and these must be well understood when planning, designing, and implementing projects:

- Changing business environment
- Increased level of competition
- Accelerated project schedules
- Unfamiliar new technologies
- New business models with integration aspects outside the control of the team
- Change-driven personal stress effects on team members and other stakeholders
- High costs for evolution of networks in a era of tight budgets
- Importance of communications is escalating a core process for project teams

Some degree of project management is needed for any project, and the more complex the project, the greater the need for the management and the more formal the processes become. Telecom projects are more complex by far than most other projects, and they are also generally larger. Both of these factors increase the demand for project management in order to enhance project success. Projects in telecom occur today in environments dealing with rapid change, and many widely different aspects of the project environment are changing as the projects proceed. Even one of these changes would be significant justification for strong management of projects. But in the electronic communications environment, many of these diverse changes are operating in parallel, so if project management is needed anywhere, it is needed desperately in today's telecommunications industry.

CHAPTER 3

PROJECT MANAGEMENT BASICS

What is a project? It is important to differentiate the project from day-to-day work. There are many sources of information on proper project management, but the best is the Project Management Institute (PMI) book of knowledge on project management called *A Guide to the Project Management Body of Knowledge,* generally referred to as the *PMBOK.* All of the material in this book is consistent with the PMI recommendations, and some definitions were taken directly from the *PMBOK.*

In order to discuss project management effectively, let us first consider the definition of a project. This is especially important since we have already seen that there is not a lot of commonality amongst the many different types of projects that occur in the electronic communications world. They are all grounded in development or application of communications technology using a network to transmit information over a distance, but since we call them all projects, they must have more in common than this. So let us start there. What is a project? The definition, taken from the *PMBOK,* is *A temporary endeavor undertaken to create a unique product, service, or result.*

Let us think this through. It is a temporary endeavor. So a project is not day-to-day work that keeps coming as time goes by, in which you routinely get some jobs, generally similar in nature to other jobs that come by, you do them, and then you go onto the next one. For example, working in a trouble reporting center could be day-to-day work. Solving one of the problems that caused someone to call in a trouble report is usually not a project. This is day-to-day work that has to be done. Trouble reports come in, and there are procedures in many cases guiding the troubleshooting, which leads as quickly as possible to a solution. There might be a need to collect some statistics as each trouble is solved, along with doing the actual troubleshooting. Generally, also, someone needs to get back to the customer with an acknowledgement that the problem has been solved. Then the worker moves on to the next trouble, which could be similar to or quite different from the one just solved. The work is generally all done by one person, perhaps with some help in some areas from colleagues, and the work generally involves only one discipline or technical area. The work is also usually completed in a relatively short time, usually within minutes or at least hours, and the ticket is closed. Resolving one trouble ticket, or even sets of trouble tickets, does not constitute doing a project. This is ongoing work.

The project is a temporary endeavor. With a project there should be a planned start date and a planned end date or due date for the output to be available, and, generally, this is the case. Sometimes, the date gets pushed back for some good reasons, such as the lack of management sign-off on the authorization in time, or delays in budget approval, or temporary unavailability of necessary people or equipment, any of which may cause delays. For whatever reason, many factors may cause the start date to move. Because of this, when we set up the schedule for our project, we will not define a start date initially. We will line up all of the necessary work in a logical order. This will give us an idea of how long the project will take, but it is generally not possible at the outset to determine the exact length of the project, as some of the activities may be influenced by constraints restricting them to specific dates or time windows. Exact timing of these constrained activities can only be determined when we know the overall timing of the project. Once the start date is determined, we can lay all of the project's required activities out on the calendar based on that. This can be done quickly once we have the logical flow of the activities and, of course, the duration of each.

Once we have all of that, we can forecast the date by which the project can finish. That date is generally extremely important, because in most projects the company, the customer, or some other stakeholder has already determined when the output is needed, so the due date is known. In fact, as the project progresses, it is not uncommon to find some request or requirement that tries to pull the completion date earlier. That request may or may not be possible to accommodate, depending on many factors that we will discuss later. For some projects, it is possible to extend the due date without many serious consequences; even so, some stakeholders will usually not be happy about this, because they have been expecting the output at a certain time. For other projects, it is not even possible to move the end date. Suppose, for example, that the project is to install the communications system for all the buildings at the Olympics. Completing the system 2 weeks after the start of the games is not merely unacceptable; it is pretty well useless. So each situation must be analyzed before any decisions concerning date changes can be made. The project team is a group of people, representing different skills and specialties, pulled together specifically to produce the deliverable by the required date. Thus, we have a temporary endeavor that occurs between the start date and the due date.

The definition goes on to state that the project is undertaken to produce a unique product, service, or result. A "product" in this context does not have to be a hard, tangible deliverable. In the telecom industry, the product delivered at the end of a project could be a new piece of equipment, a new configuration of different pieces of existing equipment such as a customer network, or a part thereof. The project could be in place to produce a new service, and while electronic communications services generally have lots of hard tangible components, the service itself is really the new concept or idea being delivered. The product could be a process, such as a new process to do something in a quicker way or with better results. The development of the process meets the definition of a project. But whatever the end deliverable is to be, it has to be unique. That is the key word here: a unique product or service or result.

This raises the question, "What do we mean by unique?" Let us attempt to describe this via some examples. One of my recent projects, for IEEE, was the development of a wireless communication engineering technologies certification. The major output of this project is a new certification program, but there are many sub-deliverables that make up the program. There is a definition of the scope of wireless communications that needs to be tested; a database of questions; an exam; IT work to allow people to register for the exam, take the exam, and get feedback on their performance; as well as a database of those who have passed, on exam performance, marketing of the program, policies to be set, strategies to map out, training guidelines to be provided to companies that want to provide training for the exam and a handbook to explain what is required to apply for the exam, and how to do so. As a related activity, the team also prepared a *Body of Knowledge* book, along the lines of the *PMBOK,* to help to describe the scope of the field. This certification program project could be said to have an output that is unique. There are other such programs in existence, such as the PMP certification program, and the IEEE Computer Society has two certification programs, but they are in software engineering, so they differ from this one. This program was developed by the IEEE Communications Society, and ComSoc does not have another similar one. There is no other certification program in existence that is the same as this one, so this program can be considered unique. The output is unique, the team is unique, and the work to develop this large program is also unique.

However, many things that are much less unique also qualify well as projects. Let us continue to look at IEEE for another example that could qualify here. IEEE runs many local meetings and hundreds of conferences, ranging in size from small local gatherings to major international events. The Communications Society alone runs over 50 conferences every year. Their two flagship conferences, ICC and Globecom, are very large, and they happen every year. Each mid year, there is an ICC, and later in the year there is a Globecom. Each ICC has a whole lot of similarities to every other ICC: they all have 54 sessions, 48 of which are technical, the rest being business applications sessions. They all have a banquet. They all have an awards luncheon. Attendees must register for the conference, and pay, either before or on arrival. Each receives a bag full of conference related material including a program, lists of the meetings, and sometimes other information such as local restaurant addresses. The conference sessions were organized around papers that were submitted as a result of an annual call for papers, reviewed by three reviewers, and then assigned to sessions by a technical program committee. And the list of similarities continues.

But the organization of one ICC is in fact a project. Although there is a great deal of parallel activity and many similar deliverables from one ICC to another, the organizers are producing, in each case, a conference that is unique. The conferences are in different locations. The technical topics are different. The structure of the sessions is different. Sometimes the technical sessions are all in a hotel. Sometimes, the people sleep in the hotel but the meetings are in a conference center. The conferences occur in different locations; they are organized by different teams; they have different technical themes, different attendees, and different organizational challenges; so each is unique. Sometimes, people call such projects repeat projects, because the teams can refer to a common organization structure for the project management, taking advantage of any planning that applies from previous such projects, yet still making new arrangements for anything that differs for their specific project.

An example from the telecom or IT world of a repeat project could be the rollout of a new application on all the computers in a company. Perhaps the company introduced one application last year and now during the next two months they are rolling out another one. One deployment is similar to another but they are different as well, because they are for different applications. Maybe they go to different end users, have different authorization levels for different people, need access or interconnectivity to different databases or technologies, and so on. So even though something is a repeat, it can still be a project, and thus it will require project management. The team must implement all the processes and techniques that we discuss for a project, and do all the things we are going to talk about for a project, but they may be able to simplify the work in some portions of the project by building the project management according to an existing PM structure that worked well for the previous projects, modified to fit the current one. The structure can then be reused each time there is a similar repeat project so that the team does not have to start from square one. They need to determine what can be reused from the management of the previous projects, and decide how the current project differs. They can then change any components of the project management structure as needed.

This gives us the definition of a project. But, in fact, there are many characteristics that are seen in projects. Let us consider some of these. For example: they operate under some significant constraints, they are performed by people, and they should be in place to meet some defined objectives. We will be talking about project objectives in Chapter 6. We look at the others here.

A characteristic of projects is that they have significant constraints. One constraint is usually the budget. Prior to the project manager and the team being recruited for the project, someone will have done an analysis, and a certain amount of budget will have been allocated for the project. The team will be quite lucky if there is a little bit of leeway in this amount, but if there is any differential between the assigned budget and the amount that is actually needed, it is rarely a large one. Often, when the project planning is completed, teams find that not enough money has been assigned to complete the project as they would like to do it. We have all seen projects that have gone over budget, and we all know that this is not well received. A good project manager wants to ensure that this does not happen.

Another constraint is often time. The due date is set based on many factors, and the work required to complete the project might not even be one of these. Even if it is considered, it is only one factor. More likely, the product or service needs to be ready to meet a market window, regardless of the work required to design and build it. Therefore, it often happens that when the planning is done, the team finds that there is not enough time available to do the project as they would prefer. We will look at many ways to collapse the time of a project, but most of these involve increasing the budget or not doing something that should be done.

Another constraint is sometimes the skills available. Or perhaps it is another resource, such as lab space, or available circuits, or test time. Not being able to get enough of the required resources creates a very difficult situation for the project team.

All of these constraints create pressure for the team. They must meet the project objectives within the start and end date window, within the budget, and within any other constraints such as the availability of people or building access on a weekend. The team has to do the work with the resources that are available.

The next characteristic of projects is that they are performed by people. Many people think that if you could just get rid of "people issues" there would be very few problems on a project. But, the people are an integral part of the project, so the people issues must be handled. The project manager is usually working with a temporary team, made up of people who may not have had prior experience working together or with the project manager. Project teams tend to be multidisciplinary, in order to provide all the skills required to develop the product or service. On some teams, there are sales people, marketing people, someone from operations, and maybe some manufacturing people, PR people, someone from purchasing, and so on, along with engineers. Few of these people truly understand the technical aspects of the project, and few people from any one of these disciplines have a good understanding of the underlying drivers behind the requirements of the other disciplines. Since the team is temporary, the people often do not know each other well. Thus, while the work progresses, the team members need to use some of their time and energy to learn about each other, and to try to understand why the other team members make recommendations and decisions that may not make sense to them. If there is not enough time and opportunity for team members to understand each other, conflicts and problems can ensue. On the other hand, if the team can take the time and use the effort required to allow people to know and understand each other, the overhead for the project might initially appear to be higher, but usually the payback will make this effort well worthwhile. The team members will be happier, the work will flow more easily, and people will learn more and come up with wonderful solutions that are more acceptable to others.

The team also needs a strong leader to set reasonable objectives for the project and provide good internal communication to ensure that all team members have the same understanding of the objectives for the project. Clearly, this understanding will enable the team members to work in the same direction to achieve the objectives, within the required dates and budget, with a product that works well and is of good quality. And in most projects, people very seriously want to meet the objectives, even though they may have different opinions about how this can be done. Making this happen is part of project management.

The accepted base of project management processes and process areas is defined by the Project Management Institute in the *PMBOK*. This organization defines 42 project management processes categorized into nine different process areas. All of these processes, in all of the process areas, can be applied to telecom projects in all of the company types described in this book. We will work through some examples and suggest some case studies as part of the material in this book. Beyond that, it will be up to the reader to take the concepts, tools, and techniques, and apply them in his or her own focus area.

This book mentions most of the processes in the nine process areas, but we will go into detail for only a few of these, the ones that are most critical for most telecom projects. The recommendations are illustrated by examples of a wide range of projects that occur in the telecommunications environment. In particular, the incumbent operating company environment will often be used as an illustration. Projects from manufacturing, newcomers to telecom, and crossovers from other utilities or environments will also receive some attention.

The process areas to be covered (see Figure 1-1) are taken from *A Guide to the Project Management Book of Knowledge* published by PMI (fourth edition, Table 3-1, p. 43). These are:

- Integration
- Project Scope Management
- Time Management
- Cost Management
- Procurement
- Risk Management
- Communications Management
- Human Resources Management
- Quality Management

Project Management, then, according to the *PMBOK,* is the application of different things, including knowledge, skills, tools, and techniques, to project activities in order to meet each stakeholder's needs and expectations. Consider what this means. In order to apply project management, we have to learn something and understand it. This produces the knowledge. We need to practice the appropriate activities or

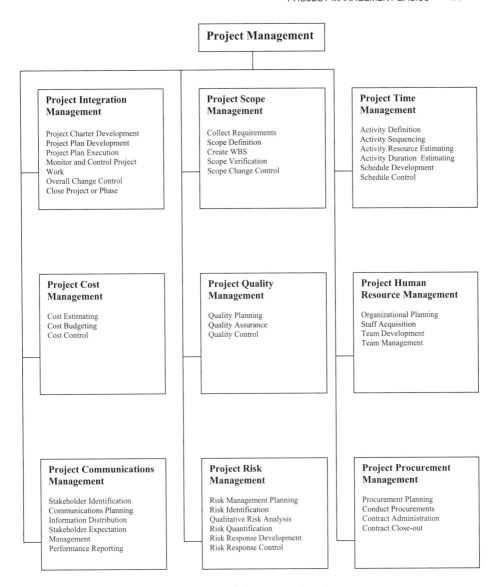

Figure 1-1. Project management knowledge areas and project management processes.

apply the appropriate techniques to develop skill. In order to effectively apply some of the knowledge or skills, we will need some tools to assist us. And, finally, we need to use the right techniques in order to effectively apply the knowledge and skills to our project. In this book, we will consider a few of these things. We will define the activities that are required to produce the deliverables that we need to produce as outputs from our project, always keeping in mind the needs and desires of people who are involved with our project.

We have to meet the needs of the project stakeholders. Stakeholders are people who have a vested interest in the project—people who care. We discuss stakeholders in Chapter 5. We need to ensure that we completely understand who might have an interest in the project so that we can manage not only our own actions, but also the fallout of anything that these people might do. Some of the stakeholders will be people who want the project to succeed, such as a customer or a senior person within the company who has something to gain from having this project succeed. Some stakeholders will be brought into the project picture because they have something that is needed for the project. And others will be people who might prefer to see the project fail. For example, with most political activities, there are proponents and opponents for an issue. The proponents will naturally support any project that furthers their side of the issue, whereas the opponents will rally against it. There can be non-supporters for any project, and these people could create problems for the project team. They have an interest in the project so they are stakeholders. And to effectively manage the project, it is important for the team to understand who these people are, what their interest is in the project, how they can be helpful or hurtful to the project, and what they might do, or be convinced to do, related to the project. Understanding all of this is part of good project management.

Generally, companies select good people as project managers and good people for the teams as well. And these teams usually work very hard to ensure that they produce what they believe is needed to successfully complete the project. With a good definition of the deliverables and the work that is needed to produce them, most teams will work diligently and produce results, and in most cases this will make the stakeholders happy. However, sometimes even this is not enough. There have been projects in which the teams have worked well, understood the deliverables and the requirements, and produced products or services that perfectly meet the defined requirements. The team did every single thing they were supposed to do, met all the project requirements, and produced the initially defined deliverable. Then, once they finished, they were dismayed to find that the customer (or even their management) said that it was not good enough or, more exactly, no longer good enough. The reaction was that the product delivered was not acceptable because the expectations of key stakeholders changed during the course of the project. At the other extreme, there have been other projects that produced outputs that were nowhere close to what they started out to produce and yet the customer was highly satisfied with the results. Again, the expectations of the customer changed as the project moved forward, and what the team produced at the end matched the expectations of the stakeholder at the end of the project. So it is quite important that the project team keep in contact with the key stakeholders, with a relationship that is close enough to ensure that they understand each other as the project progresses, and that the project work can remain synchronized with the needs for the output. This is not necessarily consistent with other aspects of project management, which require that the results be defined very clearly early in the project so that the team can acquire all the resources, including people, skills, money, and time, in order to successfully complete the project. Thus, we can start to see that project management requires a lot of juggling, and that there is quite a complex set of variables to be managed in any project, even if the project itself is fairly simple. The project team might have to not only understand, but also manage the expectations of some of the stakeholders in order to ensure that they will be satisfied with the end results that they will receive.

Project management is a science of its own or maybe more accurately, it is partly an art and partly a science. As mentioned above, there are 42 different processes that are defined by PMI grouped into nine different process areas. To manage a project, it is necessary to understand all of these processes, and to apply any that are relevant to the project. In some cases, the overhead of using many tools and techniques will be the only way that the project can be completed successfully. In other cases, this overhead needs to be minimized, with the team using only those processes and tools that will enhance the end result. There are many project-related variables in each of the process areas, resulting in a large number of variables to deal with in any project. With so many factors to track and control, it is clear that some level of focus will be needed to ensure that the right level of management is applied, and the tools and techniques used will be helpful to the team and not overly burdensome. Some of the variables are related to people, some are related to finance, and some are time or quality related. Some are directly related to the specific nature of the project, whereas others are related to projects in general. It is clear that the project manager and the team members do not have full control over all the things that they encounter as they move the project forward.

Because of the number of factors that affect projects, and the volatility of some of these, there is no guarantee that even with perfect project management you will always produce the perfect project. A project manager might do all of the things recommended here and still have a project that is not successful because there are so many variables. But studies of projects over time have shown that if you consistently apply these project management processes, the probability of success in your project goes up. In the long run, a company or organization will come out a lot better overall when project management is used effectively. Things will still go wrong, and projects will still sometimes fail, but proper project management practices will greatly increase the probability of success.

Let us look at the process areas.

INTEGRATION

Integration is putting everything together and looking at the big picture of the project as a whole. All the different aspects of the project must be considered end to end and also as they relate to each other. If we run into problems in one area, this may impact one or more other areas as well. All the aspects and their interactions must be considered in a comprehensive plan. We need to consider the life cycle for the project and we want to make sure all the different activities of the project work together effectively all the way through.

PROJECT SCOPE MANAGEMENT

Project scope management includes a number of different steps, each of which helps to build the definition and management of the scope of the project. First, the charter is created by the project's sponsor to give the project manager the high-level information as to what the project is all about. The project manager then works with the team to create first a full scope definition, and then the work breakdown structure that shows all of the deliverables and the actions required to produce them. Once the scope has been fully defined, the team can then decide whether or not they can deliver all the requirements within the project constraints. If this will not be possible, the issues can be addressed early to prevent the expenditure of resources toward a goal that cannot be reached. Once the work is underway, the team needs a mechanism for dealing with any scope issues, because in all projects someone proposes some number of changes to the agreed upon scope.

TIME MANAGEMENT

Most projects have significant time goals. Given this, it is critical that the project team have some techniques and methods to ensure that everything is delivered in full, on time, and as planned. The time management processes do this. Starting with the bottom level elements from the work breakdown structure, the team determines all of the interdependencies of these activities, and lines them up in an optimal structure for completing all the project work in the least possible time. This creates a diagram of the logical flow of the project, allowing the identification of all the dates for every deliverable and activity of the project. From this diagram, which is created to show the critical path of the project activities from the time perspective, the team can determine which activities need to be monitored closely to ensure timely completion, and which have some flexibility in completion dates.

COST MANAGEMENT

As with the time constraints, projects usually have significant restrictions on cost. Therefore, there is a strong need for processes and techniques to assist the team in defining the budget required to properly complete the project and for monitoring and controlling the spending so that all the deliverables can be produced within the funding specifications. In addition to the need for defining the resources needed, for monitoring the expenditures, and for managing these, there could also be issues related to the timing of the expenditures. If money is not available until some point in time, it may not be possible to make certain purchases or hire people before this time.

PROCUREMENT MANAGEMENT

What will the team do in house and what will be obtained from outside providers? Procurement management is the addressing of this question and following good procurement processes for anything that is to be obtained from outside the company. Both goods and services can be obtained from vendors, and in many cases there are significant legal issues to deal with such as contracts and RFPs. There will be many decisions to be made and some of these will require the advice of legal people or at least the specialized knowledge of purchasing departments whose experience includes how these things should best be managed.

RISK MANAGEMENT

No project is without risks. Sometimes project managers or teams list the things that they consider to be risky on their project, and this list is quite short. That might be fine in the very early stages, before significant planning begins, but every project has many risks. Many of these might not be significant, but still, if they are not managed properly, problems will occur. In a project such as building a new telecom service or designing a network for a large customer, there are likely to be hundreds of risks, and foreseeing these risks and putting plans in action to mitigate them can make all the difference to the success of the project. In a very stable environment, there will be fewer risks than if the environment is in turmoil, but even in a stable environment, a project is in place to do something new and different, so risks are always to be expected. In the telecom industry, the environment is anything but stable.

COMMUNICATIONS MANAGEMENT

All projects need to have effective communication, and it will be effective only if it is well managed. One perception, which does have some truth to it, is that engineers hate to deal with interpersonal communications. And another one is that engineers also dislike managing activities when they could instead "just do it," and with telecom projects, at least some of the team members will be engineers. But in a project, the right communications must occur, and they must be effective for the project to succeed. In fact, if you do something and do it really well but you do not tell any-body that you have done it, and how good it is (maybe even why it is good), nobody knows. So you do not get the credit for having done it and, perhaps, whatever was produced never even gets used. Worst of all, other team members may waste time and energy doing work that is no longer necessary. The stakeholders do not know that you did it; they might not understand what the deliverable is and they do not understand why it has the characteristics that it has, and so on. Communication is very important on projects to not only explain the services and products that are be-ing produced, and maybe how to use them, but also to report project status as the project moves forward. Regular and clear reporting to the company's senior man-agement, key customers, and other significant stakeholders is a must. In addition to reporting facts and information, managing expectations is very important, and the work of managing customer expectations is real work. Not all team members appre-ciate this, and many teams do not include time and energy to ensure that these com-munications happen, but they will take time and, thus, they will cost money. Project communication is "real work," not just that overhead that comes along with the real work. We need to put the resources into the project plan to do it and we need to make sure that somebody is assigned the job of each communication. The assigned people need to understand the job and what they are aiming for with their communi-cation. This is a very important aspect of the management of the project.

HUMAN RESOURCES MANAGEMENT

What skills do we need? Where do we get them? How is our team structured? Is the project manager the boss, and do the team members work for him or her? Or do they get borrowed from their real bosses to work on the project on a matrix basis? And then every time the real boss wants them to do something, how will they bal-ance these requirements versus those assigned them by the project manager? It is very important to get these ground rules settled and agreed with the project stake-holders. Once launched, the project manager still has many HR factors to consider. How does the project manager manage the team members within the team struc-ture? How can he or she deal with conflict? How can they be motivated? How can the project manager develop leadership in the team members, or does he/she even want to? If completing the project requires a lot of innovation, how does the project manager develop innovation? There are many human resources issues that come into play when managing a project.

QUALITY MANAGEMENT

There are a number of processes involved in managing quality, and it is well known that there is an entire science that has developed around this area, with the proliferation of programs such as Total Quality Management, Six Sigma, and so on. In projects, we need to consider some of these issues, and we need to ensure that the team understands what quality means for their project, and employs the appropriate techniques to obtain the desired quality levels. These apply to the product that the project is in place to produce and also to the management of the project itself. The quality standards must be defined before the project starts. The team must understand what these objectives are. The team then needs to be able to measure as the work progresses to determine whether things might be trending away from the quality required, and to bring them back when needed.

This covers the nine process areas, and there are project management process covering every one of those areas. This book touches on all of these, with more focus in some areas than in others.

CHAPTER 4

GETTING STARTED ON YOUR PROJECT

WHY DO PROJECTS?

In this chapter, we look at the reasons for doing projects and the initial steps in getting things ready to go. Consider the reasons that people do projects in the first place. Generally, people do projects for one of two reasons. Either they have an opportunity, such as a brilliant idea for a new service, which in the telecom area might be something like a social networking site along the lines of Facebook or an add-on application for Facebook. We will talk a little about this service shortly, to illustrate one example of a type of service that is very new in telecom, but a good model for understanding what sort of service might be a popular one. The second reason that companies decide to undertake projects is to solve a problem. In the telecom world, this could mean finding new procedures for handling trouble calls, or finding a way to bill for the new portfolio of services that are not necessarily well handled by existing billing systems. But in addition to the two standard reasons for doing projects there is a third one in telecom, and that is that projects are often undertaken to build something in response to a regulatory ruling. Thus, the motivation behind one project could be significantly different from that behind another one. Yet it is clear that in any of the cases in which there is an obvious need to produce the end result, there is probably a critical time window in which this should be available, and there will be other constraints and restrictions as well.

Consider the case of the new opportunity. In traditional telecom days, a new opportunity generally consisted of a new service or, more recently, maybe offering a new product such as a cell phone or a line of cell phones. Or, for the line side of a telco, the new opportunity might have been designing a network for a large customer where there was potential to pick up new business for the telco. In the traditional days, the new opportunity was seen as quite different from the existing offerings, and, therefore, worth the time and trouble to create a project with a well-thought-out plan and a well-stocked team. There were time issues, because for the new service, for example, some sort of market study had determined that this new offering would be popular with the customers, thus making them happy and bringing in money for the telco. The team would assess technology requirements, perhaps purchase or develop some new hardware or software, and build the service. Plans for marketing would be developed, maybe as part of the project, or sometimes in parallel with the project, and marketing materials would be prepared. Everyone involved would be trained, and much of the training might have also been part of the project. This constituted a solid project, which was generally considered to be very interesting, and which had all the features and problems discussed so far for projects.

If we look at a similar project today, we could again decide to offer a new service that has been identified as having potential for bringing value to customers and revenue for the company. This new service, though, would probably look dramatically different from the one above and from most other services that the company already offers. At least if the company is an incumbent telco this would be the case. They would have built their current business on services that were very different in nature from the one that is under consideration today. And if the company is not an incumbent telco, the service might well be similar to others that they offer, and they probably do not have to worry about the established base of telco services as the incumbent does. But they are still operating in a flexible, competitive environment. Thus, the picture of the value of the new product and how it fits into the base of knowledge of the company and its employees differs depending on which type of company is building the new service. We will address the different perspectives in the next chapter. For now, we need to look at some perspective, so let us choose that of the incumbent telco. This new service today could be something along the lines of a networking site, which could exist for social networking or for business-oriented networking. Examples of such sites are MySpace, Facebook, and Twitter. Although this has not been the sort of offering that was provided by an incumbent telco, there are many such services offered today, and more are springing up all the time. There are some similarities from one to another, but each has its own features. Let us look at one.

Let us consider Facebook as one example of a social networking site. It has seen very rapid growth, at least when compared with a typical telco service. A few years ago it was nothing, and now it is huge. It started out as a space for people, usually university level people, to post up a little bit of information about themselves and make that information available to their friends so they could keep in touch with them, know where they were, and allow their peers to keep up with what they were doing without requiring that the students put in a lot of effort sending things individually to each of the friends. The site developed with a wide base of capabilities, and today it continues to grow the sort of services that it offers to those who subscribe. They can put their pictures there. They can keep updating short statements of how they feel right now, and make comments on the world in general. They can comment on their photos or on those posted by others, even others they do not know personally. They can send greetings such as waving, throwing a sheep, bodyslamming, and so on. They can post little flags on a map showing all the places they have visited. They can have a pet, and the pet can have pets, or habitats, or clothes. They can pet, feed, and race the pets of their friends and others from around the world.

Creating such a service is a great opportunity for somebody, and of course only the imagination limits the breadth and scope of such opportunities. And usually what happens, in fact, in any company, is that people come up with great ideas, lots of great ideas, and most of those would be wonderful opportunities but the company cannot implement them all because they do not have enough resources. So they have to first consider which one of these great opportunities they can afford to take advantage of. That thinking usually happens before the project manager gets involved. Somebody somewhere within the company considers that there is a new opportunity and decides that it looks good. There are many factors they will consider, such as how much it will cost, do the skills to design and build it exist within the company, how much revenue could it bring in, will it sell, and so on. And they make a decision based on the best facts they can get as to which ones to accept and which ones to reject.

The incumbent telcos know how to do this very well for their standard types of services. However, as mentioned, something like a social networking site is very different from their knowledge areas, and the answers to these questions are not at all firm. The skills needed to design and build such a site might exist to some extent, but the technologies that would be used are changing almost daily, and it is difficult to maintain a staff with such current knowledge. The service is a very open service, which allows others to build some of the applications that interwork with the basic service. As one example, the pet application was an add-on, and some of the things that can be done with the pets were also add-ons. Such open platforms are not standard offerings for incumbent telcos and the learning curve to be able to manage these well is many years long. These services are generally offered to the end user free, with revenue coming from other sources such as advertising. This requires very different thought processes in the design of the service from what the incumbent telcos have in place already. So almost every aspect of such a new service contains some unknowns, which places the providing company in a very shaky position and dramatically increases the pressure on the project team.

With such projects, at the point at which the project manager is hired for the project, most of the benefit analysis has already been done. Someone else has already thought about what this opportunity is, and what product or service will bring success. The people who did the analysis have decided, to some level of detail, what the product looks like, and probably what they will offer as the initial components of this offering. The project manager needs to get the documentation and discussion information about their thoughts and think about what they have proposed. The project manager has to understand what he thinks they are getting as the output from the project, and why he thinks they want that. Project managers need to understand who they think the market is for this service, what they think it looks like, and how big they think it is. It is important that project managers understand when they think the service will be needed and why they want it at that date. So the first job of the project manager is to start asking a lot of questions. This is necessary not because the project manager questions the decisions or the proposals being made, but because the PM really needs to have a very good grasp of all of this information in order to make the right decisions as the project moves forward.

Designing a customer network today might look somewhat similar to designing a customer network ten or fifteen years ago. But today, in addition to all the problems that telco engineers have always faced, there are added problems. First, the technologies available are constantly changing so any design must be able to use and cope with any new technologies on the scene. And the product must be something that will compete well against a whole raft of competitors, many of whom could well be companies that the provider does not know. So the project team is working in a much more volatile environment than in the past, an environment with far more competition. The project *must* be well designed and executed or the company could well lose all the money and effort put into it. Of course, if the product is a new line of cell phones, all of the above issues apply in spades, and now the company is working with something that has become a commodity. And telcos, along with their suppliers, are still building their expertise in this area.

Thus, the first reason for undertaking projects is to take advantage of an opportunity. The project manager might think of that opportunity himself or herself with the resulting need to then have to do all that business analysis and to sell the concept to the company. Or the idea might have originated elsewhere in the company, and someone else might have done the analysis, which is provided to the project manager when he or she takes on the project. If the project manager and, later, the team members believe in the possibilities for this service or product, people work really hard because they know that this is something that makes their company money. And usually these projects are fun as well. The Wireless Certification project which is mentioned from time to time in this book is one of these. It is something that the entire team believes will bring good value to the individuals who take the exam, as well as to industry. In addition, it is something that is going to bring in money for IEEE Communications Society. It brings value to business, and value to business is what we have been trying to do. We believe that. That is what you want to create in the minds of all the team members on any project. If it is possible to get everybody signed up so that they really believe in the project, they will do all the things needed to make it happen. This is one of the strongest factors in success.

Another reason for undertaking a project is to find and build the solution to a problem. For example, there might be a problem with inefficiency and the project is to find and implement a way to improve trouble handling. Maybe this is happening in an internal call center that takes calls from people calling in with network or service problems. According to the practice, the next available agent picks up the call, analyses the situation, and then hands the problem to whomever is best equipped to handle it. In some cases, maybe the agent who takes the call is also a technician qualified to fix certain problems, and he decides that he can handle it. What the trouble handler might do then is to ensure that he or she understands the reported problem, perhaps do some testing to see if he can recreate the problem, and then come up with a new design or fix for the problem. When the trouble has been cleared, someone needs to report back to the customer that the problem has been solved, perhaps ask the customer to try out the solution, and then close the trouble ticket. Perhaps the problem is that customers think it is taking too long to close the tickets, or management thinks it is costing too much to close the tickets, or perhaps the people who handle the troubles think they are closed whereas the customers are still having a problem. This would be a communications issue. The first order for the project team is to clarify just what the actual problem is, so that they can start to look for the best solution. The project is to find a better way to manage the call center, or the trouble handling. This is quite a legitimate project as well.

And finally, some projects are initiated neither because the company sees an opportunity nor because the company has identified a problem needing a solution. In telecom, as in many industries, projects can also result because the regulator or the government decrees that something must be done. An example could be the implementation of cellular local number portability. None of the companies would have likely initiated that themselves but it is in place in North America today because regulators told the mobile call providers that they must implement this. Any of these projects can be turned into opportunities, or can be used to also fix existing problems. But they initially are undertaken because of a decree.

THE REQUIREMENTS

Projects can be undertaken for different reasons. They exist within companies that are vastly different in their organization and business goals. The projects themselves can be similar or they can be radically different from one another. However, there are some general requirements that are common to all projects.

First, there will always be a requirement for producing the end result that drove the creation of the project in the first place. The word for this in project management is the scope. Every project has a scope, and this scope clearly describes the product to be produced, whether the product of the project is an actual product or a service. The scope also includes all of the management that will be done to manage the project successfully. This makes it necessary to clearly define the full scope of the project and the management of the project, and it only makes sense to do this early in the project. Not only must all the elements of the scope be provided, but they must be provided at the required level of quality. Therefore, it is also critical to clearly define the expected quality for all the deliverables, and even the activities. If this is defined and understood by those who are involved, they have the opportunity to work to meet the requirements. If the requirements are not clearly communicated, they might be met, either by fluke, or because everyone happened to understand them anyway. But given the fact that we have people with very different backgrounds, views, and goals, it is not a good idea to decide that they all have the same understanding of all of the requirements for scope and quality. It is important to define these. We will address this later.

Second, there is usually some date by which the end results are needed, and possibly even dates by which some interim results are needed as well. This is the time requirement. Almost every project has time requirements and, of course, there is a lot of management required to ensure that these can be met, given what we have already discussed about the environments in which projects are undertaken and the nature of projects.

Third, there are also usually cost requirements. It is very rare for anyone to assign a project with the comment that cost is no object. So, again, it is very important to manage to meet the cost requirements.

But consider the connectivity amongst these requirements. They are interrelated. It is usually possible to make up time when things start to run behind by either hiring more people or having people work overtime. Of course, both of these come with a cost. Hiring additional people will affect the budget, and if these people are not as skilled or knowledgeable as the initial team, we might also suffer some quality issues. Overtime might also have a budgetary impact, but even if people work free overtime they will become tired and this raises the possibility of quality issues. And in the case of either being behind schedule or being over budget, someone is likely to consider bringing things back in line by dropping some of the scope. So a problem in one area is likely to also impact another of these areas. And, further, if all three (or four, if you separate quality out as a fourth requirement) are set at the beginning of the project with no leeway on any of these, then the project will almost certainly fail, because the team will be hard pressed to correct any issues in any one area without impacting another area.

WHAT HAPPENS BEFORE THE TEAM ARRIVES?

As mentioned above, before any team is assembled, someone in the company has usually completed a fairly comprehensive business analysis of the project and has decided to go ahead. Because every company has many possible courses of action, there are always quite a few suggestions on the table, any of which could be excellent. Companies usually assess these, often with some very involved processes and techniques. Sometimes, two levels of assessment are done, with the first assessment being quite high level, not addressing many details, just to determine whether or not it is worth putting in additional time and effort to make a more informed decision about moving forward with the idea. The second stage would then be more involved, and sometimes this second stage is actually a project in itself. In any case, once these assessments have been done there is information about the project, including what is expected as the desired outcome, why this is important, why it would be of value to do this, and many other factors related to time, cost, and benefit of the project. It is wise for every project manager to obtain as much of this information as possible to use in making plans and decisions for going forward. It is quite possible that directions may change as the project planning proceeds, and even that these changes might be good. But it is best that the decisions for the changes be made consciously and in an informed manner. Then the team can ensure that no one is expecting something different from their end results, and that whatever they do produce will maximize the benefits to their company.

Also, before the project manager is brought on board, the company has to decide how the project will be funded, and the mechanisms for this funding. Once the budget to which the project work will be charged has been identified, the sponsor has been selected, because the sponsor is the internal senior person who is paying for the project.

One of the first things that the sponsor needs to do to get the project going is to prepare the project charter. The charter is a document that gives a high-level description of the project. It should be kept very short, even though it contains a lot of key information, because it becomes the first communication document for the project. The recommendation for the charter is that it be three or four pages in length. It contains an overview of the project scope that can be discussed amongst all the potential players. The sponsor will use this information to explain the project to a potential project manager. Once the PM has accepted the position, he can also use this document to recruit the team members, possibly also sharing it with the functional managers who will give up their people to create the project team. It might even be shared in some cases with suppliers as well. Other information that might be found in a charter can include the due date for the project and maybe for some of the interim deliverables; a list of the skills needed, maybe with the names of some of the key desired people; some indications of the budget requirements; perhaps some processes that might be followed; some risks that could be encountered; some of the assumptions that the company is making; and maybe some of the key stakeholders. Obviously, if all of this information is to be included in a three page document, there will not be much information included in any of the areas. Other documents must be created later in the project to document more of the details.

It is clear that the charter is a communication document. It will be the first entry in the project plan. But it also serves another purpose as well. Once the sponsor and the project manager agree on the concepts presented here, they sign the document and this authorizes the PM to start working on the project. An account is established to which the project expenses can be charged, and the PM can start to hire the team and begin the work.

SETTING PROJECT OBJECTIVES

This topic will be the topic of the next chapter, so we will not address it here, except to say that at least some high-level objectives will be set prior to the team being put in place. Although some of these might be project-related objectives, many will probably be business-related objectives. It is important that the team understand these objectives, and also that they understand which ones they can meet and which are to be met by others.

GENERAL DESCRIPTION, SKILLS ANALYSIS, STAKEHOLDERS, AND RISK ANALYSIS

By this point, we have a general description of the product to be produced and some of the structure for the project management. The project manager has enough information to start assessing the skills and resource requirements for the project. When he understands the skills needed, he can determine who should be hired for the project team and approach the functional managers with his requests.

When the team members have been assembled, they can start to assess the information on the project scope and work on the details of exactly what this means. They can start thinking about the deliverables, what is feasible in their environment, and how the project requirements can be met.

Some of the initial considerations need to be a consideration of the stakeholders, and some initial risk analysis. In the next chapter, we will talk about the stakeholders, and in Chapter 9 we look at the management of the risks.

CHAPTER 5

WHO IS INVOLVED?

In this chapter we consider some of the people who will be involved with the project. We have already mentioned the project manager, and he or she will be the key person on the project. We have also mentioned the sponsor, and in this chapter we will further clarify the sponsor's role. Then we will discuss the rest of the project stakeholders.

PROJECT SPONSOR

The sponsor, as per the *PMBOK,* and as mentioned in the previous chapter, is the person who pays for the project or represents those who are making the investment. It is usually necessary for the sponsor to be someone within the company that is doing the project, because the project team needs a lot of ongoing assistance from the sponsor. The role of the sponsor cannot be effectively played by someone outside the company, even though it is quite possible that the project costs might be covered by someone outside the company. First, the sponsor provides the funding for the project. Although it is certainly true that there are many projects that are done directly for a customer, and that the customer is charged for all of the project expenses, this money rarely goes directly from the customer to the project team. The money is paid by the customer to the company providing the product or service, and the company pays the project team via charges to an internal account code. This account code is generally assigned to some manager who is charged with managing the expenditures, probably through the project manager. This internal manager of the account is then the sponsor.

But even aside from the funding, the team has other expectations of the sponsor. If the sponsor is in senior management, he or she will have inside information regarding the corporate direction, and probably also the relative importance of the project within the company. Only one project can be the most important one to the company and, conversely, one has to be the least important. Obviously, when the project managers for these two projects approach functional managers to recruit their people, unless there are some very specific skill fits, the manager of the first project will have a much easier time getting the best people, whereas the second PM will face some challenges. If the PM is aware of this before the negotiations begin, he or she can better prepare. The sponsor can share this knowledge. The sponsor can help the PM to understand how his project fits into the corporate strategic direction and how it fits with the reward system of the company. Armed with this knowledge, the PM can make better decisions on behalf of the company as the project progresses, and can ensure that the value the project brings will be something that is important to the company. The sponsor may also be the person in the room when the management of the company discusses potential budget cuts. If the sponsor has a good understanding of the importance of the project, and also of the current status at the time of the discussions, plus the upcoming needs of the team, he can defend the project in the budget discussions and protect the needed funds. He can also help the other senior managers understand the importance of the project.

In addition to support at the senior levels, the project manager may need some assistance from others within the company. The PM does not have control over the managers in the other business lines and functions, but often the project needs something from these others. The best approach is for the PM is to work with the others, sharing information and giving as much as he can to the others to convince them of the benefits of assisting him. If the PM has good people skills, this will work in most cases, but from time to time there will be conflicts that the PM will be unable to resolve; in such cases, it can be very useful to have a sponsor who can and will step in to assist.

The sponsor, being more senior, is also in a position to give more meaningful rewards to the team members. If someone very senior does something special for a team member, it will generally mean more than getting the same thing directly from his immediate supervisor. Sponsors will often help with rewards.

STAKEHOLDERS

Stakeholders include anyone with an interest in the project. Very early in the project planning, it is important to identify all the stakeholders and also what their interests are in the project. When the team understands the potential for the involvement of the stakeholders, the project can be designed to take advantage of any support that might be forthcoming and to minimize any potential disruption. But in order to do this effectively, it is important that there be a good understanding of what this potential is. Hence the need to take a little time to identify the list of the stakeholders for a specific project and to identify what they might be looking for related to the project. If the project is a repeat project, such as the design of another new long distance service or the implementation of a second or third call center, the lists from the previous projects can be used as starter lists and modified as appropriate for the new project.

Stakeholders, including people such as your boss or your project team, can be supportive. These are the people that the PM would expect to help to get anything needed for the delivery of the project within the requirements. Stakeholders can be people who will be helpful to you, even though they might not have much reason outside your project requests to know much about the project. Stakeholders may be within your company, associates of the customer, or external interest groups whose association with the project may not initially be obvious. Stakeholders could include the functional manager from the department down the hall, whose people you need to borrow to do the project. This person does not have any specific interest in the project of his own, so he does not necessarily want to give the project manager his best person go to work on the project, but it will be tremendously helpful if he does. Because he has nothing to gain directly from the output of the project, he will have to be convinced to give up a good person, because he will have to do his own work without that person while the person is assigned to the project work. For some projects, assignment of people could be full time for an extended period of time, so unless there is some benefit coming back to this manager, he will not be well motivated to let his person work on the project. For this reason, it is necessary for a company that frequently does projects to have a project-oriented culture in which a manager's providing staff in support of an important project is expected and rewarded behavior.

Stakeholders also include people who may be against the project, such as a team with another project that is competing for resources with the PM's project. These people may even engage in "political" destructive behavior to make their project look more important for the company than the one selected. They might find a way to make the PM's project looks like a loser so senior management will reallocate the money to them. This is the sort of behavior that theoretically never happens in a well-managed company but unfortunately is all too common.

So we could have stakeholders with views from one end of the support scale to the other, and all of these people can potentially impact the project. Therefore, one of the first things a project team should do is to sit down and say, "What's my project? How big is it? What does it look like? What do I produce? What's the objective? And who cares?" Create a list of stakeholders. It does not really take long to do this. Ask many questions, including:

Who cares if we do this project?
Who cares about the outcome of the project?
Why do they care?

It is important to know why they care because if you know what their interests are you can sometimes align the activities of the project to benefit the stakeholder, thereby getting the best chance of their cooperation in aid of the project. In order to convince other managers to do things that will advance your project, it will be important to convince them why work on your project is more important than what they would otherwise be doing, which is obviously a challenge. Or, if the stakeholder happens to be someone who will work against your project, the team must be able to figure out what the stakeholder could do to hurt the project, and then think of ways in which they can prevent or mitigate the damage. So we need to know who the stakeholders are and we need to design our project in such a way that we can take advantage of those things that can help us and that we can be ready for, and deal with the ones that can hurt us. We need to understand the stakeholders. And our goal in the whole thing is to meet their needs and expectation.

MANAGEMENT

One special stakeholder is the management within the corporation or the organization that is doing the project. Support of corporate management is required at the highest levels to establish and maintain the project-oriented culture that is necessary for a company to accomplish things by setting up and executing projects. All managers in the company undertaking projects must have a clear understanding of the priorities of project activities versus ordinary day-to-day work if the project manager is to get the needed support.

The team has expectations of the management and of course, the management will have expectations of the team. It is important that the team determine what these expectations are, and although many of these might seem to be overhead that is not productive in moving the project forward, the management requirements should be met, unless the team can convince management that for this specific project some of the standard or usual procedures should be set aside in order to obtain greater benefit from the project. One overhead that should not be scrapped is two-way communication with management. It is very important that the team keep management informed of progress and issues, and vice versa.

FUNCTIONAL MANAGERS

Another group of stakeholders that must be seriously considered is the set of functional managers whose people are working on the project. This is a group that is often neglected once the project team has been assembled. But since they have given something valuable to the project, and they might be needed to do this again on future projects, it is wise to consider these people as members of the extended team. When the team receives any project symbols, such as hats or t-shirts, why not order a few extra and give them to the functional managers as well. Invite them to the parties!

THE TYPE OF COMPANY HOSTING THE PROJECT

Again, the material covered in this chapter to this point applies to all projects. But there are things that are specifically telecom related that we need to also consider. We mentioned the scope of the telecom industry, which includes end users, service providers, distributors, manufacturers, and others. Now let us look at their different perspectives on telecom projects. We will look at the projects from the following perspectives:

- Incumbent established carriers
- Newcomers to the telecom field
- Established companies in other areas who are new entrants to telecom
- Manufacturers

Of course, any of these companies might have projects of any nature, and for those projects that do not affect the customer, the main difference in the project environment from one company to another is the internal corporate culture. This is more a function of the style and views of the corporate management than the type of company, although, as a general rule, most established incumbents, especially those that are large or work in a regulated environment, will have many people who are risk averse and are comfortable working with quite a lot of bureaucratic overhead, whereas smaller companies and start-ups are much more likely to be freer in their approach to business and to the way in which things are implemented.

In the telecom environment, the incumbent carriers and, generally, also companies such as cable and power companies that are already established in another area, tend to be very procedural, exhibiting a great deal of bureaucracy and not much tolerance for risk. They have a long history of learning what works for them in business, and they believe they know well how to manage in their market space. Of course, as we have seen, all of this is changing. But the company brings this baggage with them, positive or negative, when projects are undertaken. In addition, such companies have huge investments in equipment, technologies, existing services, customer base, practices, procedures, employees, employee training and support, and so on, and may tend to overestimate the value of their legacy technologies and services.

Newcomers to the telecom field, aside from those moving over from other utilities, tend to fall more into the category of smaller companies or start-ups. Many of these start-ups, including companies such as Google or Second Life, are no longer very small. But in many cases they still retain that entrepreneurial spirit and approach to business that helped them make the inroads that have brought their success. They do not bring heavy bureaucracy with them. But what is happening to some such companies is that they are being acquired by larger companies, and their larger, more established owners might have a business philosophy more dominated by established processes and procedures, and often a very different approach to the market. Projects in the more entrepreneurial companies may occur in a freer environment. This can mean freedom from procedural overhead; it can also mean freedom to mess up really spectacularly. In any case, the companies, new and old, are facing the same rapidly changing market requirements.

Within an incumbent carrier, there is usually quite a different perspective on the company, the products, and the directions in those departments usually known as line departments compared to that of the departments known as staff departments. The line departments work directly with the customers. They have sales people; customer systems engineers; service managers who implement networks and services, fix them, and handle problems; plus people who do other customer-related functions such as manage customer-related projects or program customer-dedicated switches. On the other hand, staff groups evaluate technologies that might be used to provide the underlying network and services, such as packet-switched networks versus circuit-switched networks, frequency allocations, and so on. The focus of the groups is quite different, even though both are needed to provide the full scope of customer service and products. It is important that both of these areas keep in mind the views and considerations of the other, in order to optimize the overall end product for the customer.

Manufacturers will have a wide range of corporate behaviors, as many manufacturers are established long-time providers to the telecom industry, whereas others are smaller, more nimble start-ups with more leading-edge and innovative products. Their biggest stakeholders from a business perspective are the carriers or other service providers, while the biggest stakeholders for the service providers are the end users.

As should be clear, the focus of one of these companies in terms of products and in terms of the needs to the stakeholders is quite different from that of another. The type of products or services might differ from one company to another. The incumbents are providing networks and services that are heavily reliant on the networks. The manufacturers are providing the products and technologies that ride on or use the networks. Some are trying to protect their large sunk investments, whereas others are just trying to capture a portion of the market at whatever the cost might be. Therefore, the project objectives for a service provider that is an incumbent carrier will have some parallels to those of a service provider that is trying to break into the market, but many of the objectives that relate to their overall corporate objectives will be quite different from one company to another. The people on the teams need to be aware of this, because what might be appropriate for a project in one such company is not necessary a good way to manage in the other.

CHAPTER 6

SETTING BUSINESS AND PROJECT OBJECTIVES

In the charter, there will no doubt be some high-level objectives for the project. These might be for the project as a whole or for different aspects of the project, such as objectives for cost, time, quality of the end product, or even the management of the project.

Some objectives will have been predetermined by the initial analysis that resulted in the decision to undertake the project, but this is a business analysis that determines whether or not completing the project will enhance the business position of the company. Because of this, the considerations undertaken beforehand are usually business considerations and not necessarily project related at all. As a result, the high-level objectives that are set during this feasibility analysis are very likely to be business objectives, which may not directly translate into specific project objectives.

So what is the difference?

Business objectives are targets which, if achieved, will enhance the business position of the company. Some examples of business objectives could include:

- Design and implement a new Internet service to generate over $15M profit over the next five years.
- Implement new processes for selection of positive new business opportunities that will produce three net positive services over the next two years.
- Overlay the existing cellular network with a 3G network capable of carrying twice the traffic volume currently carried by the corporate cellular networks by November 2012.

- Design and develop terminating equipment to provide WiMax service to four specific U.S. cities, and coordinate with the service provider to have the service operational by March 2012.

These are all good business objectives, as each seems to be such that the results would place the company in a better business position in some way. The first two of these are objectives for the business, but not for the project, whereas the second two could well be project objectives as well. Let us look at the reason for this. With the last two objectives, we see that the aim is to provide something that could well be provided by the implementation of a specific project. Of course, since we do not have all the details of the project, it is possible that the projects do not include all the work and deliverables required to allow the meeting of these objectives. So we should first ensure that the objectives can actually be met within the project work. If that is the case, then we could say that the last two objectives are good objectives for the project team.

In the case of the first two objectives, though, unless the projects run for five years in the first case or two years in the second case, these objectives cannot be met during the time of the project. And, probably in both cases, the people who would do much of the work to help meet the objectives are not part of the project team. For instance, in the first case, the sales force is needed to generate the sales that will bring in the revenue. If these people are members of the project team, then it is reasonable to have an objective in which they participate. But if the project manager has no influence over the work of the sales team and, in addition, that work will happen long after the project closes, not only can the project evaluation not be completed, but the project team has little opportunity to ensure that they meet the objective since they are not responsible for a large portion of the work needed for the objective to be met. Thus, the first two are good as business objectives, but not as objectives for the project team. The PM and the team should work to ensure that they have the capability to be able to meet any objectives assigned to the project.

SMART OBJECTIVES

In different references, there are various definitions of what constitutes a goal and what constitutes an objective. For this book, we use a well-known and frequently used technique that uses statements that are SMART as objectives. By this we mean that the statement is specific, measurable, achievable, reasonable, and time bound. In addition, the statement must clearly be something that the team must or wants to achieve, rather than just statements of facts or descriptions of things that happened.

Let us consider what is really meant by statements being SMART. In order to really understand this, it is best to analyze each of the components of the requirements for the statements. We can look at what is meant by each separately.

1. Specific—The objective should clearly state what is to be done or produced so that there is no mistaking what is needed.
2. Measurable—You should be able to measure whether you are meeting the objectives or not.
3. Achievable—Are the objectives you set achievable and attainable?
4. Realistic—Can you realistically achieve the objectives with the resources you have?
5. Time—When do you want to achieve the set objectives?

Specific

An objective statement must be specific. This means that the objective must specify what the team is to achieve via the work on the project. An objective is specific if it is stated to describe something that is clear, concrete, detailed, focused, and well defined. In other words, if more than one person reads the statement, each would have the same understanding of what the team is aiming for. In this way, we avoid misunderstandings and prevent the possibility of different people working hard for different things, simply because they thought that what they were aiming for was the objective. A specific objective is results and action oriented, and it clearly states what the required outcome is to be.

Measurable

Objectives must be measurable and it must be clear to all stakeholders how the results are to be measured, lest different people use different measures, leading to conflicting verdicts on whether or not the objective has been met. The measurement is the standard that the company will use to determine whether the objective is being met, but when setting measures to be used, be careful. People will work to meet the set goals as they are measured. It is necessary to ensure that what is being measured is actually the desired behavior. For example, if the desire is to design a process for trouble handling that clears tickets more quickly than they are currently cleared, be specific about what is meant by rapid clearance. If the only thing measured is the time from ticket opening to ticket closing, this will generate behavior that closes tickets quickly, but not necessarily behavior that actually finds and implements true solutions before the tickets are closed.

Achievable

Objectives need to be achievable. There is not much point in setting an objective if the team will be unable to meet it, because we risk demotivating people, as an unachievable goal is a guaranteed failure. For example, if the completion of the work to enable the achieving of the objective is too far in the future, it will be difficult to generate excitement and sustain the motivation. It is fine to set what is known as a stretch goal, which is something that will make people work a bit harder and smarter than usual. This might even be more motivating but it should not be so extended so far as to not be reachable.

Realistic

Realistic is not the same as achievable, so objectives must also be realistic. To be realistic, an objective needs to be something that this team can reach with the skills and resources available to them. When an objective is achievable, it could be that it would be difficult or even not realistic for a given team, at a given time. If so, it might be necessary to either change the objective or to find additional resources for the team in order to make the objective also realistic.

Time-Bound

Finally, it is important that objectives be time-bound. An objective is time-bound if there is a deadline specified for the achievement of the objective. Deadlines need to be both achievable and realistic. If no time is set, then the work can drag on, as there is reduced motivation to bring the project to conclusion. Without a time goal, the team will feel less urgency to execute the tasks. Setting time limits creates the urgency and prompts action.

When the objective is stated in such a way that it meets all five of these criteria, the team and the stakeholders will be more easily able to determine how to meet it and whether or not it has been met.

CHAPTER 7

WHAT IS TO BE INCLUDED?

The clear comprehensive definition of the scope of a project is one of the biggest keys to success. And unfortunately, in telecom as in many industries, too many projects are started before this clear definition has been worked out, and sometimes the full definition never does completely emerge.

In this chapter, we look at the development of the full scope description, starting with the Charter, which we have already discussed, then developing the scope description, scope management processes, and the work breakdown structure.

BUILDING THE CHARTER

If all goes according to expected procedure, the project sponsor will have developed the project Charter and used it to hire the project manager. But in many cases, in both established service providers and small start-ups, sponsors and other executives do not have time to draft documents for projects. Instead, they discuss the concepts and the requirements with someone more junior, which for a project would normally be the project manager, leaving the PM to build the Charter. This could be a good opportunity for the PM, because it allows him to draft the Charter to fit well with his understanding of the requirements. If he can produce a document that is acceptable to both himself and the sponsor, then the initial project documentation will be in place.

In order to create a project Charter, it is necessary to learn as much about the project requirements as possible. The writer must describe exactly what the project is in place to produce. The product description will not be detailed, but it must be clear and give a good idea of what will be in place by the end of the project.

There should be some indication of the money available for the project, and if there are time constraints related to the funding, these should also be included. It may be necessary during the project build stages to justify the expenditures by showing some significant return, either financial or some other form. It is best to build the detailed plan with an understanding of any such requirements.

The Charter should also define any major constraints that exist. These could be physical constraints or logical. They might be budgetary constraints if there is a hard limit on the dollars that can be spent. The budget per se is not a constraint; it is actually an objective, albeit an objective that it is advisable to meet. But in the case in which there is a hard limit to the amount that can be spent, this limit must be listed as a constraint rather than as an objective.

In some cases, it is useful to include information about people, companies, or groups that will be impacted by the product under development. If this information is known, the people should be identified as well as some information about the impact. The team can then consider how the project can be designed to ensure that the desired results can be achieved.

If the project is a line organization project for a specific customer, it is best to get as much information as possible at the early stage about the customer's expectations. Any needs of other known stakeholders can also be incorporated early in the project design.

If the PM takes the time to look into all these areas early, the team will be in a better position to plan the project effectively.

If, as is often the case, the PM is directed to develop the Charter for the sponsor, he then needs to discuss the details with the sponsor, to ensure they are in agreement on the constraints and the positions of the stakeholders. This discussion helps to clarify what the project scope is and secures executive sign-up for what needs to be done. Once the goals and deliverables are agreed on, the PM can start to think about who is going to play the different roles. A sample project Charter is shown in Figure 7-1.

SCOPE DESCRIPTION

Once the Charter has been accepted, a full description of the project scope can be developed. This is documented in the project scope statement or scope description. The scope statement is a narrative description of the project, building on the initial information in the Charter and elaborating on it to clarify both the product and the project.

Since this is a very complete description of the product to be produced and the work that is to be done to get there, it will be a long document. It is recommended that this document be as complete as needed to ensure that everyone who needs information gets a clear understanding of whatever is needed. Therefore, depending on the complexity of the project, the number of participating groups, and the number of interested stakeholders, this document might be 300 pages long and it might use charts, graphs, pictures, diagrams, brand names, and standards to show the in-

PROJECT CHARTER

Project Overview

Project Title: **Google Drive**
Prepared by: P. M. Getsaround   Contact info: PMG@interactiveserv.com
Project sponsor: F. Godmother   Contact info: FG@plutocrat.com
Client contact: A. Traveller   Contact info: anyt@public.com

Project Objective:

What will be accomplished by completing this project? Please specify the reasons for undertaking the project, the benefits that will be obtained, and the time frames within which the benefits will be realized.

- Establish partnership agreement by the end of January 2012, with Google, to combine Google Earth in a real-time service offering over our mobile network.
- Establish relationship with Google to allow future service builds, with an agreement by end February 2012 to plan two additional services in 2009.

Project Scope

Provide a brief description of the product or service to be produced. Give information about the methodology to be used.

In order to offer benefit to our current customers and to attract new ones, we propose to deliver a series of mobile services that use GPS and Google capabilities, initially Google Earth capabilities, integrated with capabilities of our network. The initial service will allow mobile customers to view areas en route via their mobile phones, and to query specific information such as travel distances. In order to enable the development of the first service and subsequent services, it is necessary to develop an agreement with Google. We prefer to build this partnership and a subsequent relationship rather than arranging a right to purchase/use the Google capabilities to facilitate true interaction and also to pave the way to possible future joint ventures.

Project Deliverables

List all of the major project deliverables.

- Service specification for Google Drive
- Project plan
- Analysis of benefits for Google
- Agreement for Google Drive service
- Basis for cooperation for future interactive service development

Inclusions/Exclusions

Specify key items to be included in the project, and those items that will not be included in the project or the end product.   *continues*

Figure 7-1. Sample project Charter.

Inclusions	Exclusions
Meetings to discuss relationship parameters	Consumer market research
Entertainment for Google executives	Network capacity and loading studies
Full evaluation of mutual benefits	Google Drive service development
Project planning and management	Licensing our service to other mobile operators
Impact assessment of graphics display on mobile phone screens	

Assumptions/Risks

1. Mobile phone screen display will provide enough resolution to facilitate use of the service while driving.
2. Another network provider might offer more benefits earlier.
3. We can obtain enough setup time with Google executives to allow us to present benefits to them.
4. We will be able to negotiate a deal that will allow both companies excellent prognosis for profit on Google Drive.
5. Antitrust suit by competitor to prevent the joint venture.

Budget

In 2008 R&D budget:	$1.7M
In 2009 R&D budget:	$0.2M
In 2008 negotiation budget:	$1.5M
In 2009 negotiation budget:	$0.6M

Constraints

List any known project or activity constraints related to the scope, timing, cost, or quality of the project.

- Any agreement signed must allow our company to maintain our standard profit margins.
- Negotiations must follow completely ethical practices.
- R&D portion of the budget must not be exceeded.

Additional Customer Requirements

None noted.

Success Measures

Specify the success indicators for the project, with potential requirements for each.

- Draft initial agreement developed by early December with mid-level management agreement from both companies
- Agreement allows for interaction and mobility in real time.
- Profit sharing levels are commensurate with level of risk and contribution to the service.

Figure 7-1. *Continued.*

Date project required by: March 1, 2012

Department approval from: James Deanne   Dept: Marketing   Tel. #: 8-1592
                           Randy Fletcher   Dept: Legal   Tel. #: 6-4567
                           Dig Calcul   Dept: Finance   Tel. #: 4-5678

Prepared by: P. M. Getsaround. Contact info: PMG@interactiveserv.com

Client Signature   _____

Project Manager   _____

Project Sponsor   _____

                   _____

                   _____

Figure 7-1. *Continued.*

formation as clearly as possible. Anyone who reads this should be able to fully grasp the nature of the product and the project. So the scope statement has to be very clear and detailed.

At this point, it is also important to specify how the team members will know that they have produced what they should be aiming for. What are the stakeholders looking for from the project? It is important to understand how they think the end product will look, and what they are going to measure to determine that they have what they want. If the team does really well in meeting the schedule but the stakeholders are focused primarily on the budget, the stakeholders might be unhappy even though the schedule was met. In the scope statement, it is important to define the success criteria.

Properly defining the scope involves viewing the project from a number of different perspectives. This process should be undertaken by the project manager with assistance from as many members of the project team as possible. The initial step is to review the charter, ensuring that there is a full and common understanding of the contents. Next, the team should consider the context within which the project is being undertaken. As we have discussed, in the telecom environment three or four different companies might all be undertaking a project aimed at producing the same end result, but the project can look very different in each company. This is especially true for similar projects undertaken in two different companies with varying cultures and business models. In telecom, factors such as the existing equipment and other investments, the customer base and the corporate approach to the market are usually company-specific. So two projects which might appear identical to an outsider may look very different to the teams.

For project teams in telecom, it is not enough to undertake the analysis and detailed definition of the product by the team; additional information needs to be added to the picture early in the planning stages. The team needs to identify as many as possible of the project stakeholders, some of whom may not be obvious at first glance. This takes some thought but should not take a large amount of time if there are team members familiar with the different aspects of the project environment and what people or organizations may be affected by the running of the project. Once there is a list of the stakeholders, it is important to identify their potential needs. This will give the team the opportunity to look at the project from the perspective of these stakeholders. They will all have needs, and they will be looking to the project to help meet these needs. If the project does not do so, they may not accept or support the project. This analysis will give the team the opportunity to consider ways of designing the project to be positive to the stakeholders, who will generally have differing or contradictory needs and priorities. As a result it will typically not be possible to meet all the needs of all the stakeholders. The project manager needs to be able to juggle these differing requirements as the project unfolds to best satisfy the range of stakeholders.

Before proceeding too far, the project manager should ensure that the project has the appropriate approvals. Sometimes, the sign-up of the project sponsor does not ensure that all the necessary internal approvals are in place, and if these approvals are not forthcoming, work will have been expended for nothing. It is crucial to ensure that executive approval has been granted by those executives who are truly empowered to make the decision; external approvals, from outside the corporate hierarchy, can be even more challenging.

In one project with which I had experience, the sales team had sold a new offering to a large customer. This offering would use telecommunications to change the way they offered their services to their clients. The client was extremely supportive of the change at the upper management level, because management could see the possibilities for much greater client acceptance. However, this was a large organization, with a number of unions to deal with, and a huge line organization dealing with their customers. Since the solution being built for the client was integrated with its internal systems, operations and processes, there was a strong need for the project team to work closely with the client middle management in the definition of the project and product scope. However, when the project team approached their customer contacts, they were rebuffed, and no real communication of the required details occurred. Without the ability to obtain client details, the team was not able to clearly define the project scope. The cause of the problem was a perception on the part of the client at the working level that their jobs would be impacted by the project. They were not willing to support the definition of something that was potentially a threat to them. Eventually, this project was cancelled because the client management and the project team were unable to convince the client working level of the benefits (to them) of this project. In this case, the identified client stakeholders did not really have the ability to make the call for their company. This sort of issue should have been resolved early in the project.

From the project charter, and including the results of the stakeholder analysis, the team can create a narrative scope statement. According to the *PMBOK*® Guide, creating this scope statement is part of scope planning. The statement will include a full summary of the project. With the scope statement in place, the team should then build the scope management plan, followed by the work breakdown structure.

The scope statement will contain information in each of the areas covered in the Charter, with any appropriate additions or clarifications, as well as any additional information that is needed to clearly describe the project.

The project scope statement should contain at least the following types of information.

a. Business Need

This description is based on the information in the Charter, but further details should be included. The business need should describe the opportunity the project will exploit or explain the problem to be solved. The opportunity should be quantified in some way in terms of a measurable desired outcome. A problem should be defined in terms of the gap between "the desired state" and "the current state."

b. Project Justification

The fact that there is an opportunity that would bring advantage to the company or a problem that needs to be solved does not justify undertaking a specific solution, or ensure that the company can afford to do something, or even do anything. State the reason that this project should be undertaken. Explain why the company should expend resources to meet this need. The project justification provides rationale to justify the expenditure, and to justify the undertaking of this project rather than other projects that would meet other business needs. The rationale could be an increase in revenue, an improvement in customer service that will help in specified ways to maintain customers, an improved visibility, and so on. There should be enough information provided to allow management to understand the benefits of this project and to compare these benefits to those of other projects competing for the same resources. It will be beneficial in selling the specific proposal if the rationale is directly related to the corporate goals and mandate.

c. Product Description

A project is undertaken to provide some deliverable or product. The outcome of the project might not be a product per se, but a new service or a process implementation. However, from a project management perspective, there must be something that is to be produced. The product description provides a brief narrative description of the "product." It is quite acceptable to refer to other documents that might include some of the details of this description, as long as those reading this document are also able to obtain the reference material.

d. Project Deliverables

Every project can be decomposed into a small number of high-level deliverables. The scope statement must list these. There should be at least one deliverable related to project management. At the highest level, this can be just "project plan," or "project management." Because this project management is going to be part of doing the project, the scope statement is a good place to introduce it. If it is not introduced at this point, it must be included in the next step, the work breakdown structure. The deliverables should be first expressed as large, high-level deliverables. They describe the "what" of the project. Listing these allows people to comprehend the components of what the project is producing.

e. Included/Not Included
The scope statement should clearly describe the items that will be included in the product, and the project. At a minimum, this should be an itemized list, but if needed, further description should also appear. More importantly, the statement should also specify what will *not* be included in the project. This will give further clarification to the product and the project, and possibly provide flags for some stakeholders. It is unusually easier to resolve such potential issues early.

f. Project Objectives
Project objectives can relate to every aspect of the project, including the overall results, the timing, the cost, the scope or the quality of the work and the deliverables.

Overall project objectives might be taken directly from the Charter, or they might be new objectives that have appeared during the initial project management discussions.

In the scope statement, often in a section relating to timing, the team should list the major milestones that management will use to measure the success of the project. There will probably be some objectives related to some of these, and those objectives should be stated in this section. For each of these, a time frame should be specified. If some of these milestones also pose project constraints, those should be mentioned under constraints. However, constraints are different from objectives, so any one item can appear in at most one of the two sections. Objectives should be consistent with the relevant project selection criteria and with the assumptions and the constraints contained in the scope statement itself. Objectives related to project costs should be consistent with project selection criteria for profitability and/or growth.

Some companies are very focused on quality, and in these companies most people are generally fairly sensitized to the need to define objectives for quality. In good project management, these objectives are set for the project, covering at least the major deliverables. In fact, it is advisable to set quality objectives for all activities during the detailed planning stages. In the scope statement, the objectives for quality should be specified for the project itself and for the high level deliverables. The team should specify the performance expectations that must be satisfied in order for the project to be considered successful. These objectives must be consistent with any quality program in place within the company, and at times they must be consistent with customer quality programs as well. These should also be consistent with the project selection criteria and with the project justification.

g. Project Constraints

A constraint is a significant limitation to any portion of the project. All projects have constraints, and it is important to identify and communicate these early. This section should identify any limits that cannot be crossed by the project team when providing the deliverables.

There is a difference between a constraint and an objective. A constraint is a criterion that must be met; an objective is one that it is desirable to meet. Therefore, no statement can appear in both classifications.

h. Project Assumptions and Risks

When planning a project, there will be many things that are not known during the planning stage, and yet in order to prepare a complete plan, the team must make assumptions, remember that assumptions have been made, and build plans on this foundation. During the planning process, as information continues to be developed, it is necessary to check back and reverify that the assumptions are still valid. Anything that the team has used as true information in order to facilitate the planning of the project is an assumption until it can be proven. If an assumption turns out to be false, the overall picture of the project will be different from the one that drove the plan. Some parts of the plan may no longer be feasible, requiring significant changes. In the scope statement, it is important to identify any significant assumptions that will be used in the planning.

Any known project risks should also be listed in the scope statement, so that the project can be designed to handle them. The list of risks will be extended later in the project, as described in Chapter 10.

i. Success Measures

The statement should clearly define how project success will be measured. Objective, measurable criteria used to determine and measure project success should be described. Also include information on when the measures will be taken and, possibly, how or by whom.

The scope statement should be prepared by the entire project team, to the extent possible. Since this statement is prepared early in the planning cycle, it can happen that some of the team members are not yet assigned. In this case, it might be necessary to prepare the scope statement with only a partial team, but then it is wise to try to bring in expertise in all areas of work to ensure that nothing is left out or planned incorrectly. Preparing the scope statement can be time-consuming. For example, the author once worked on a project to create some new features for a long distance service, where the team of over 30 people spent a full day developing the scope statement for the project. After the statement was developed, they moved forward on the development of the work breakdown structure, only to find that they could not complete this structure because the scope statement had not been properly developed. After taking another day to replan the scope, the team found that the WBS followed quite nicely.

In fact, this particular project was one in a series of related projects. Each of these projects was built upon a foundation that was established by the previous one in the series. Hence, the first project created the core features of a long distance service, which went into operation once it was developed and accepted. The second project added additional features to the service created in the first project. Of course, since the first project had produced an operational service that was now live, the team for the second project was presented with new information that had previously been unknown, including customer reaction to the new service, problems with equipment, network, and processes. In addition, as each project was completed, since the timing was critical, there were some components that were deferred from one project to the subsequent one, leaving the teams for the third and fourth projects with bigger requirement sets than they had initially planned. This structuring of projects to allow one to build on features provided by an earlier project is one typical way telecom projects are handled. The advantage is that some service can be provided early, and also any customer reactions can be factored into the planning for the subsequent project.

The completed scope statement *must* be approved by the project manager, and *should* be approved by the key stakeholders. The team should take the time to share the information, if not the statement itself, with all the key stakeholders to ensure that there is common understanding of the project and the product to be produced.

Let us look at the information that can be included in a scope description. I have decided to use as an example the scope statement for a large project I have been managing for IEEE Communications Society (Figure 7-2). Although this is not a telco project, it is definitely a project in telecommunications and a very interesting one at that. The project is the creation of a certification program for professionals working with wireless engineering technology. Although it is not likely that many companies would build a certification program, there are quite a few company-specific examples of such programs today. This project should be understandable and relevant to most people in telecommunications, so it can serve as a good example. The full scope statement for this project is quite extensive, with many pages of information, so the scope statement example in Figure 7-2 is a somewhat shortened version that still illustrates the information needed.

SCOPE MANAGEMENT PLAN

Defining the scope of the project is important but it is only the first step. Once there is an agreement on the scope, project plans will be built to ensure that this scope will be produced. The schedule, the budget, and the resource allocations will all be built based on the scope. All subsequent project planning and implementation is based on a clear understanding of the project scope, so it is clear that changes to this scope will impede the ability of the team to finish the project within all the require-

PROJECT SCOPE STATEMENT
Wireless Engineering Certification Program

Business Need

The Wireless Communications Engineering Technologies (WCET) Certification is an exam voluntarily undertaken by an individual. The exam is constructed, administered, and updated by an independent third party (IEEE Communication Society), strictly following or exceeding well-established best practices. Prerequisites for taking the exam include both a recognized undergraduate degree and professional experience. Extensive research has validated the traditional value proposition of certification programs for individuals and for industry. Certification has a significant value for those wanting to stay current or looking to advance, and it can be a tool to filter job candidates, assist in initial hiring, and reduce on-the-job training requirements. The WCET certification aligns well with the evolving needs of multiple stakeholders, including industry, government, and many countries, by following the advice of a diverse Industry Advisory Board. The certification process will facilitate, identify, and promote a worldwide work force with the knowledge and competency expected by their employers. A robust promotion and marketing program will attract those who have expertise in other technology areas and want to move into wireless, and those in the field who wish to stay competent in a rapidly evolving and internationally competitive set of technologies.

Project Justification

IEEE today provides value to academics and because approximately half of the IEEE members are from industry, industry also sees value in the IEEE Communications Society (ComSoc) products. But ComSoc would like to provide better value to industry, and WCET does this by saving hiring managers time in assessing potential new hires or internal transfers from nonwireless areas to highly skilled wireless positions. WCET will provide IEEE a means to better define its role in engineering education beyond its traditional focus on outreach. In addition, WCET is a new product, and thus it can increase revenue through:

- Application and examination fees
- Study guides sales (e.g., practice exams)
- Sales of other IEEE and ComSoc publications thatcould be used to prepare for the examination
- Training program enrolments (future, if such programs are developed)

WCET fulfills the desire for ComSoc to produce new, nontraditional products and services that will be valued by the members and society. In addition to the direct benefits to hiring managers of assisting them to find qualified candidates for difficult to fill positions, WCET provides a single product that will help industry
continues

Figure 7-2. Sample project scope statement.

managers to learn of the value of IEEE and ComSoc. Offering such a new program will create considerable additional advertising, which will increase the visibility and prestige of IEEE and increase potential membership numbers, especially within the younger, early-in-career demographic. Another pull-through revenue then will be potential revenue increases related to potential membership increases.

Since other IEEE Societies will potentially decide to offer certification programs, documentation of the development of this early program will provide a model to use to create additional future certification projects.

Taking into account the anticipated direct program-related revenues, including projected exam fees, practice exam sales, and sales of the *WEBOK,* the entire WCET program will be completely profitable and in the black starting in 2010, with original start up funds being repaid in 2013. Considering the cost and complexity of the build, plus the potential for ongoing long-term revenue, this is quite reasonable.

Product Description

Develop and launch a program to confer upon applicants a Certificate of Wireless Communications Professional (WCP) upon passing the WCET examination.

WCET certification will show that the holder has proven extensive knowledge of the wireless engineering communication technologies' body of knowledge and, by maintaining their Wireless Communication Professional (WCP) title through recertification, has demonstrated the desire to remain informed and educated about any new developments in the wireless field.

The certification will be available in any country around the world. The target audience is professionals with an engineering degree and a minimum of three years experience in wireless.

The project will develop all components required to allow people to learn about, prepare for, apply for, establish an exam time appointment, and write the exam. The project includes the initial assessment of the likelihood of acceptance of a certification program in wireless and the development of the technical scope that needs to be included, specifying both the knowledge that is needed as well as the tasks that the successful candidate should be able to perform. Since the certification is intended to bring value to industry, the tasks must be applied using the knowledge needed by industry. The lists of knowledge and tasks will be assessed by at least 1000 industry people who will comment on the usefulness of each statement. Only areas that they deem to be important will be used to build the exam. A database of at least 600 questions will be developed, and from this the exams and practice exams will be built. Only the first three exam sessions will be developed during the project, with questions being retired after each exam offering and replaced by new questions.

The project will also produce a Book of Knowledge that will describe the field of wireless communications engineering. Although this book will not be associat-

Figure 7-2. *Continued.*

ed with the exam, it will be based on the same scope as the exam, and it is likely that some candidates will use this as a guide in preparing for the exam.

The project will not produce any training for people to use to prepare for the exam. However, the project will produce training guidelines that can be used by any company or university wishing to provide training. The team will post information about training that will be available from any of these sources.

The project will include the preparation and implementation of a marketing program that will include news releases, presentations at 10 significant wireless conferences, announcements to be distributed at 20 additional conferences, promotional seats for selected participants, presentations to volunteers for the Communications Society around the world, and a website.

Since some outside funding has been obtained, the project will also include regular status reports to the funders.

The project also includes IT work, including the capability for people to register for the exam; a secure vendor with international sites to allow people to apply for, pay for, and write the exam near their home locations; records of the successful candidates; and integration of this information with existing membership systems.

A handbook will be prepared that will provide information about the candidate requirements and all processes associated with application, preparation, and writing of the exam.

The program will also provide certificates for those who pass the exam, indicating their successful status.

Project Deliverables

- Market assessment, job analysis study
- Project Management plan and organizational structure
- Definition of scope of wireless communications. Identify areas of responsibility and specific knowledge and tasks within those areas that would be expected in the skill set of a engineering professional with 3 to 5 years of wireless experience.
- Final test specifications
- Databank of questions
- Exam and standard setting
- Candidate handbook
- *Wireless Engineering Body of Knowledge*—book describing the knowledge areas
- Launch of application window
- Launch of exam window
- Presentation of first certification
- Deployment of maintenance phase
- Marketing program   *continues*

Figure 7-2. *Continued.*

Included/Not Included

Included: The WCET is a certificate of Wireless Communication Engineering Technologies; all other communication specialties are out of scope.

The target candidate for the WCET certification will be a wireless engineer who holds a degree or certificate from an accredited university or school (accredited in the country in which the institution is located) and who has at least three years of graduate level education or professional wireless communication engineering experience.

Not Included: Skill sets commonly learned outside of the experience level of the targeted candidate will not be included in the certification.

Project Objectives

- Include in the exam only testing of knowledge and tasks important to industry as identified by industry professionals who review the lists.
- Examine at least 50 candidates in the first window, and 200 in each of the successive windows two per year, in 2009 and 2010.
- Create all material via volunteer manpower, using no more than two part-time individual contractors in 2008 and 2009 plus two providers who are expert in certification creation and exam distribution.
- Establish an industry advisory board consisting of at least 10 senior industry managers before the end of 2008.
- Produce all deliverables within the budget of $1.9M.

Project Constraints

- To obtain ANSI accreditation, ComSoc cannot directly create a training program; a separate effort would need to be undertaken by IEEE directly.
- Exam will be limited to wireless communication engineering.

Project Assumptions and Risks

Assumptions:

- Majority of project participation will be from volunteers.
- Industry will support certification by sending people to take the exam.
- Successful candidates will find new positions in wireless, either within their current company or in a new company.

Risks:

- Some candidates may find the fee for the certification too high versus their level of income, limiting the candidate pool.
- There is an inherent risk in developing both a market and a product at the same time. The product must be creditable enough to create the market in advance of product availability

Figure 7-2. *Continued.*

Success Measures

Success will be measured by the number of sales of the Examination, Practice Examinations, and *WEBOK*. Numbers will be calculated by the outside examination service vendor and supplied to ComSoc within three weeks following the close of each examination window. The actual number will be compared to projected sales, and the WCET marketing and informational programs will be adjusted according to any variance from the projected sales numbers.

    The initial proposal specified anticipated sales numbers for the exam, the *WEBOK,* and the practice exams in 2008, 2009, and 2010. The project will be deemed successful when these measures are reached. The initial proposal gave June of 2010 as the closing date for the project, so the numbers include only the first exam cycle in 2010.

NOTE: although this sample is based on an actual real life project, many of the details have been altered in this sample scope statement. Anyone looking for true information about this project needs to contact IEEE Communications Society.

Figure 7-2. *Continued.*

ments. If anything is added, additional time and money will be needed, because the schedule and the budget cannot accommodate the additional work. The more changes that are attempted, and the more significant the changes, the bigger the impact will be on the time, the expense, and possibly also the quality of the project and its deliverables. Changes in scope are very common in projects delivering new products and services, and such changes are almost always additions to rather than reductions in scope. The impact of proposed scope changes needs to be carefully analyzed, and if the changes are undertaken, it must be with a clear understanding on the part of all stakeholders of what the time, money, and quality impacts will be.

So, along with the actual scope statement, we also need a scope management policy. This policy must be clearly described, communicated, and implemented in order to assist the team in meeting all the project requirements. The first step of this policy is to clearly and fully define the scope, which the team follows up by monitoring and controlling the ongoing project activities. In a changing environment, however, there is much more to scope management than this.

The big project problem is "scope creep." Change requests may occur for various reasons, including missed items in project definition or planning, new opportunities that surface during the time of the project, or even unavailability of needed deliverables from another project. On any project, many people, including the team members themselves, will propose changes and additions to the initially defined scope. In most cases, these are wonderful ideas, which would produce a better product or project if they were implemented. But the original scope definition was used to create the project budget and schedule, and these will be impacted by any accepted change. If the number of changes, or the size of the changes, are minimal, the project goals can probably still be met. But at some point, every project will hit a breakpoint beyond which the project team can no longer be successful because of the new requirements. Therefore, it is best to implement a change management process from the beginning and to apply it to every proposed change, even when the impact of a change appears to be minimal. If the overhead required to analyze the request and decide and communicate the decision might actually take more time than incorporating the change, some of these changes might be incorporated automatically to save time. But even some of these proposals might be put through the process to ensure that everyone understands the need to protect the project.

The change management process should clearly outline all the steps and information needed to ensure that each proposal can be handled to the benefit of the project and its stakeholders. These things would include:

- Identification of who is entitled to submit change requests (anyone, only certain stakeholders, etc.)
- Clarification of what sort of information must be submitted to the team
- Identification of any dates beyond which change requests will not be considered
- Possible recipients of the requests, and information on how the recipients will handle them
- The length of time within which the team will respond to a request
- Any additional requirements the team might place on the submitter or others involved

But probably the most important information that needs to be spelled out in the process description is that if a positive decision is made (i.e., the change request is granted), the team will require changes to the budget, schedule, or other aspects of the project before they can incorporate the change. The PM is wise to obtain agreement from key stakeholders early in the project regarding this policy so that a subsequent request for resources will not come as a surprise.

Most of the changes that will be proposed during the run of a project are changes that are actually necessary, or at least advisable, as opposed to just nice to have, and, therefore, the assessment groups will want to include them if possible. But, the problem of meeting the project objectives still exists. The team needs to ensure, before accepting any request, that the required resources will be available to them to allow the required changes to the project plan.

WHERE WILL THE RESOURCES COME FROM?

In some cases, the project manager will be advised by the sponsor or management to use resources already available to the project. To do this will require that the resources be reallocated from the original plan to the newly proposed deliverables. Sometimes, in order to get one piece of functionality, the client is willing to give up another. But it is important, before doing this reallocation, that the PM get approval from those who will be impacted by the loss of the initial items in the plan to drop or reduce the impacted items on the project. If these people are still expecting that they will receive the initially planned items, the team will still not succeed, because there will not be enough resources to do both.

Another solution to the problem of needing additional resources is often presented. Sometimes, PMs are told to use the project contingency for scope changes. This is definitely not acceptable. Contingency is included to cover "known unknowns," which are the risks defined during the project planning phase, and will be needed for this purpose. In fact, we will see that since the technique for incorporating contingency does not include the full required amount for any of the risks, it is quite possible that there is not be enough contingency to even cover the risks, let alone for scope changes. Never use contingency for anything other than its purpose—to cover the known unknowns. For scope changes, the PM needs to obtain additional resources, which might be obtained via an agreement to relax schedule or expense requirements.

Assessing project change requests will take time. The initial project plan, in the project management section, must include the time needed for the project the team to assess change requests and replan when necessary. If any requests are accepted, there will be an additional requirement to negotiate for the additional resources that will be needed. This time must also be taken into account and included in the project schedule. Project management work is real work, involving real resources and time. These activities must also appear in the work breakdown structure, which we will discuss next.

Another consideration is the approval of project changes. It is important that the approval be in written form, acknowledging the required change and the resulting project impacts, and that it comes from the appropriate stakeholders.

THE WORK BREAKDOWN STRUCTURE

At this point, the scope and the scope management plans have been defined. Now we want to build a structure for the project, as this will more clearly define the project activities and deliverables. This structure will become the basis for the project schedule, budget, and resource requirements, and is called the work breakdown structure or WBS. The WBS identifies all project deliverables and activities. The WBS helps the PM and the project team to avoid leaving tasks out of the project; if the team is going to work on anything, it has to be included in the WBS. The WBS also ensures that there will be no double accounting for work, because another WBS rule requires that every deliverable and activity can be included in the WBS only once.

At the top level of the WBS, the team identifies the high-level project deliverables. There is no need to put any specific deliverable or type of deliverable at any particular level of the WBS; each team should place items in the way that makes the most sense to that team. But at the top level, it is best to have a reasonable number of broad deliverables, with only the most significant deliverables appearing at that level. From there, they can be broken down, with smaller subdeliverables included at the lower levels.

At the bottom level of the WBS, the deliverables become actions, and at this level it is necessary to include a verb in the description to explain the action to be taken.

Another rule for the creation of the WBS is that the elements of the full structure must integrated with the project as a whole. This means that when any one deliverable is broken into its components, all the components must be included in those shown below. Another way to say this is to say that the children must completely add up to the parent.

Since *everything* must be included, if you plan to do anything to manage the project—assess change requests, negotiate for more money, monitor activities, and so on—all of these must appear in the WBS. The WBS is a breakdown of the project into deliverable-oriented groupings. These deliverables eventually turn into action items and when that happens, the bottom has been reached.

The purpose of the WBS is to assess *what* is to be produced by the project and how this will be done. The structure must include everything that is included in the project, whether it be something for the project—in fact, all the project management must be included—or something to create the product. The WBS shows all of this—the *what* of the project. But only the *what*. There is no time element in the WBS and no cost element. These can be created later, using the WBS information as a basis to build both the schedule and the budget. But the WBS itself is not sufficient to build these.

One more rule that should be used in building a WBS is that there should not be any of what I call "single children." When the project is decomposed, it is possible that at some point a deliverable might be decomposed into a single deliverable. If there appears to be only one element below any other element, there must be a problem. Either there are additional subdeliverables making up the upper one, and these need to be included as well, or the lower element is in fact the same as the upper box. If so, then there is no need for the lower element, as only one is needed.

Once we have completed the WBS, all the project activities are defined. The PM can then assign people, time, dollars, and so on to each one. It is the bottom-level elements of the WBS that I am now calling activities. These activities must meet certain criteria. They must be:

- Assignable
- Independent
- Measurable
- Schedulable

- Budgetable
- of suitable size for monitoring

Assignable. Each activity must be such that it can be assigned to one person or group. If the work spans more than one unit, further breakdown is required.

Independent. Each activity must be independent of the others to facilitate management of the items.

Measurable. The way in which the activity completion will be measured should be defined so that stakeholders can determine whether or not the activities have been completed satisfactorily.

Schedulable. The activities will be used to build the project schedule, so it must be possible to assign a start and end date to each.

Budgetable. The amount needed for the project budget will be determined by assigning costs to the bottom-level elements. These can then be added to determine the cost of each deliverable and the cost of the full project.

Suitable size. Although there is no rule about what size the activities might be, and one activity could well be considerably larger or smaller than another, the size of the bottom-level elements must be suitable for monitoring. The appropriate length will depend on the project, on the PM, on the team, and on how critical items are.

The WBS can be in chart form, along the lines of an organization chart, as shown in Figure 7-3, or it can use the same format as the table of contents for a book. The WBS in Figure 7-3 is not complete, but some of the deliverables have been decom-

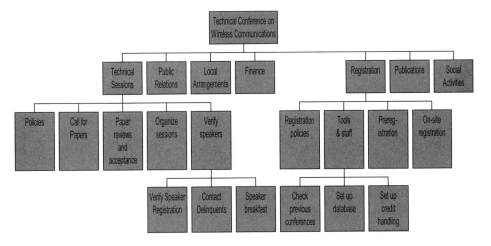

Figure 7-3. Work breakdown structure, second example.

posed to illustrate the technique. This is a WBS that I use sometimes in my courses. The outline below is a second example, showing another format that can be used for a WBS.

0. Customer Network Design for Handy Dandy Shoes
   1. Existing Network Assessment
      1.1 Collect information on design of current network
      1.2 Prepare network design drawings
      1.3 Run performance tests on current facilities
      1.4 Compile performance statistics on this network
      1.5 Identify any problem areas
   2. Requirements Development
      2.1 Determine problems with current network as seen by client
      2.2 Understand any changes client wants to make to performance parameters
      2.3 Determine the time at which client will decide on a network vendor
      2.4 Manage any problematic customer timing expectations
      2.5 Understand any changes client wants to make to his service capacity requirements
      2.6 Understand client's growth plans and requirements by location
   3. Network Design
      3.1 Determine appropriate services to propose to client to meet his requirements
      3.2 Configure the overall architecture of client network showing the services
      3.3 Run performance programs using client usage data to show expected performance of the proposed network
      3.4 Prepare professional network layout diagrams showing network and performance
   4. Sales Cycle
      4.1 Prepare presentation showing service changes proposed and reasons for them
      4.2 Prepare presentation showing network changes proposed, reasons for them, and benefits
      4.3 Provide time lines to customer for proposal
      4.4 Provide time lines to customer for rollout of new services and network
      4.6 Prepare presentation showing all financial and performance benefits client will attain in total with the new proposal
      4.7 Arrange customer appreciation day for client, during which the presentation will be made, in a quasisocial setting
   5. Project Management
      5.1 Determine risks of retaining this customer with new design
      5.2 Determine budget required to prepare and present all required information, including management of the project

5.3 Determine time lines for completing all project activities
5.4 Determine contingencies in time and money to ensure success in obtaining this contract
5.5 Monitor project timelines
5.6 Monitor project spending versus expected expenditures
5.7 Prepare and share standards for the network design
5.8 Prepare and share format and standards for the customer presentation
5.9 Identify all customer communications, with standards, mechanisms, and time frames
5.10 Identify major internal communications, with standards, mechanisms, and time frames

This WBS is fairly simplistic because it has only three levels, including the overall project at the top level. There are many additional things that could be included. What might some of these be? Is there anything missing that should be there?

Even at the bottom level, elements can still be broken down further, but this is not necessary for this WBS. Elsewhere in the project documentation, the team can provide any needed additional details.

When the WBS is complete, we will then move to the next steps, which are:

- Add duration and dependencies so we can build the logic network
- Add the calendar to the logic network to give the schedule
- Add resource names to each activity
- Add dollars to each activity

Once these have been done, we can calculate the budget and the schedule. We can also then plot out the cash flow, which might, in turn, influence changes in the schedule. We will discuss the next steps listed in later chapters.

CHAPTER 8

GOING OUTSIDE THE COMPANY FOR PRODUCTS AND SERVICES

Many projects can be completed totally in-house, using people and other resources already resident within the company. For these projects, this chapter is not needed. We need to use the procurement processes when we need to go outside the company, either to purchase something or to hire people or companies with specific skills to complete some portion of the project.

The project planning team needs to first assess whether or not it is advisable or necessary to look outside the company for any part of the project work or deliverables. In some cases, it will be clear that the expertise or material is not available in-house. In others, a make-or-buy decision may be included as part of the project work. This is all part of procurement planning. We could be looking at vendors to purchase goods that are needed to complete the project, or we could be looking at vendors and/or contractors to complete work that they can do better or cheaper than we could in-house. In either case, we would use the procurement processes and we need to start by deciding exactly what we want. The end result will be a plan specifying what will be obtained from outside, and what processes will be used in order to obtain this. In this chapter, we review some of the tools that are used in procurement or, as this is often called in telcos, purchasing. We will not only discuss the tools for procurement, but we will also consider the implications of using these.

The processes recommended in the *PMBOK* for procurement management include procurement planning, solicitation planning, solicitation, source selection, contract administration, and contract closeout. Instead of discussing the processes per se, we will talk about the tools that are used, which should give a good picture of what is required of the project team.

DEFINITION OF REQUIREMENTS

In order to prepare for solicitation, which is the process of obtaining information about what is available from outside to meet our needs, we need to first be clear about exactly what it is that we are looking for. Once we have decided what to go outside for, we need to prepare the specifications to clearly describe what we want.

SOLICITATION

There are many tools that can be used for solicitation, but probably the three most common are request for information (RFI), request for Proposal (RFP), and request for quote (RFQ). We address each of these here.

RFI

An RFI is a request for information. This is exactly what it sounds like—a request to potential or possible vendors for information on what they might provide to meet our requirements.

This document can be issued to as many or as few potential suppliers as desired, and it is acceptable to ask for any information except for a quote on the price. If a price quote is requested, this document becomes either an RFP or an RFQ, and, as such, it is subject to all the implications of those documents.

The issuance of an RFI does not imply that the requestor intends to buy any-thing. This is truly a request for information. Because of this, some vendors may not respond to an RFI. If the supplier is very interested in getting the business from the buyer, he will probably respond. But if he thinks that the buyer is just searching for information that he will then use internally, rather than looking for something to purchase, he will not be likely to respond. If the response will require much work or time, some vendors will not respond. In fact, many smaller companies may not have the manpower to create custom responses to RFIs and also do their paid work.

The preparation of an RFI can be as detailed and as complex as that for an RFP or RFQ, if the specifications are known for the desired product or service. Sometimes, the RFI is more general because the buyer is truly trying to determine what products or services could possibly be available, and he will tweak his requirements based on information gleaned from the responses to the RFI. In other cases, the buyer knows what he needs but does not know what is available or who provides such products or services. In these cases, the detailed specifications or requirements could accompany the RFI so that the responses will be more focused on the specific needs and, thus, the potential vendors for selection can be more effectively narrowed.

So the responses to an RFI can be quite varied. In one case in which the author was looking for a billing system for a new long distance service, an RFI was issued to about 10 vendors and five of these responded. Three of the companies did not even initiate dialogue. Two others did not send any actual response, although one of these asked to be included if there were to be any RFP in the future, and the last one disappeared after asking a few initial questions. This left five companies that did respond. Two of these sent material that is known as boilerplate, which is preprepared information customized by adding the name of the buyer and, perhaps, specifics related to some of the requirements. A third company directed the buyer to their website, suggesting that all the required information was available there. Another prepared a wonderful customized reply which clearly addressed each of the requirements and also presented other benefits of their system. And the last vendor sent a CD with a model of their billing system, which was quickly passed around to all the potential users to get their reactions. As this shows, responses can be quite varied.

In this case, the vendors were suspicious that the telco would not select any billing system, even though, if one were to be selected or requested, each wanted to be considered. Why would a telco decide not to buy any system, when they were willing to take the time to create, issue, and evaluate responses for an RFI? The main reason would be that the customers of the telco want their billing information to come on one single bill, and this is extremely difficult to do if some of the charges come from a separate billing system. So the telco will incorporate this billing into the existing system(s) if at all possible. In the meantime, since this incorporation is very difficult and time-consuming to achieve they might well look for a separate system that could much more easily and elegantly handle the charges for the new service, just in case it might be needed.

An RFI can be sent to any number of vendors and it can even be issued publicly via a website, bulletin board, or newspaper. The wider the distribution the greater the likelihood that a possible vendor will be found. But, of course, the more responses that come in, the more time that will be required to assess these responses. So in many cases people try to put some limitation on the possible responses.

The main purpose of the RFI, of course, is to determine who offers products or services that meet the requirements of the project. In some situations, it is necessary, or at least advisable, to issue an RFP or RFQ. The results of the RFI then will be used to create the list of vendors to whom the second request can be sent. And if there is no need for a second request, perhaps the vendor can be selected from these responses. If so, negotiations can start immediately to obtain the products or services, provided they fully meet all requirements at the best possible price.

As mentioned, there is no requirement that the company issuing an RFI ever buy anything. On the other hand, there is no requirement that a company responding to an RFI actually provide the material, work, or service proposed, although it would be highly unusual for someone to respond and then decide not to follow through.

There is a cost to an RFI, in addition to the cost of preparing the requirements, issuing the document, and assessing the responses. Perhaps the biggest cost in a project environment is the time required to do all of this. It generally takes more than a day or two to prepare the requirements and get this list approved. It may not take long, then, to prepare the RFI, but the vendors cannot respond immediately, especially if there is any level of customization needed for the product or service. Usually, there is a window of some weeks for the vendors to respond. Once the responses are in, the buyer needs to assess them, and this could also take some time, especially if these are complex or there are many of them. It is not unusual to take a full month from the time that the preparation starts till the results are complete in the RFI process.

So, although the information is very useful, the time required might cause many to bypass this process.

Generally, there are responses to an RFI that is issued. If it should happen that no responses are forthcoming, the buyer will need to either change the requirements or work through contacts and research to find some specific companies to contact directly about the need.

RFP

An RFP is a request for proposal. This is used when the buyer is serious about his request, and skills are required to provide the desired items. Thus, this tool is used when the requirement is some professional or technical work, when some design is required, and so on. If someone were to want some custom software, an RFP might be issued, and the proposals offered to provide this could provide very different products. This is the sort of thing that telcos might request, or the request might be for some type of equipment engineered to provide specific features, requirements, or service levels. If the customer is looking for a network with certain capabilities, this could also be requested via an RFP. In fact, telcos often have a group of people who are specifically charged with the job of responding to such requests. These people assess each request, and then collect all the required components of the response from the different groups of experts. This group is also expert at deciding which questions should be asked to the vendors, and at finding good ways to obtain the answers to these questions. We will return to the reasons for this later.

The writing of an RFP must be done carefully. The requirements must be very clearly defined and clearly described so that the responses will be useful to the buyer. If the requirements are not complete, or maybe worse, incorrect, then compliant bidders will be providing solutions to the wrong problem, and these will not be useful to the buyer. The basic requirements will be technical: description of the product or service desired with all size, brand, performance, standards, and so on that the buyer wants to have. The requirements may be specified in detail, covering such aspects as just mentioned, but it is also possible to specify the requirements in terms of the end use of the product or service, leaving the design of the solution to the vendor. In addition to these product-related requirements, there will be other requirements, such as the timing for the submission and, perhaps, the method of submission or inclusion of information about the bidding company. The buyer needs to ensure that all such requirements are also clearly spelled out.

An RFP, when issued, has some significant legal implications, so this document should never be issued without first working with the company's legal department. The issuing of this document signals that the buyer is planning to buy. Because of this, suppliers take more serious notice of RFPs than they do of RFIs. But because of this, it is difficult for a company to back out of the RFP if they find that they do not want to buy after all. This could happen because they discover at some point that the list of requirements does not actually specify what they need. The list could possibly be corrected via an addendum if the omission is caught early in the process and if the change is not significant. It could happen because some new regulation changes their needs, or because some new technology changes their requirements. It could happen because they find that something new has become available with which they can work much better than whatever was in their original plans. Most of these are very valid reasons to stop, from the perspective of the buyer. But the vendors, once they begin their response, have invested time and money in building their responses, and they will not be happy about doing this for nothing. The longer the buyer waits to cancel the RFP, the stronger the reaction will be. And with an RFP, it is possible for the vendors to take successful legal action in many cases. So the document and the situation need to be well thought out before the issue. And it is always wise, if there is any suspicion that the buyer might want to refuse all proposals, that legal words to this effect be included in the request. Of course, the inclusion of such wording will impact the likelihood that some vendors will reply seriously.

Another condition that some buyers might wish to include, particularly governments who generally operate on the principle of accepting the lowest priced proposal, is that the buyer reserves the right to accept the proposal they deem to best meet their requirements, even if this is not the lowest priced.

But the legal potential does not end here. When there is an RFP open, it is possible, as mentioned above, for any vendor to request additional clarification or information. The buyer may not wish to narrow the request but, usually, the buyer wants the suppliers to have as much information as possible so that he has a better possibility of getting a good proposal. So the buyer generally wants to provide any needed information. This can be done, as long as it is done properly. If any information is to be given to one of the bidders, that same information must be given to all potential bidders, so that they all have the same chance to meet the requirements. So, if you issue an RFP, and one of the bidders asks a question, you answer this by issuing a notice to all those to whom the RFP was sent, giving them the question and your response.

Sometimes, companies hold bidders meetings at which any potential bidder can ask any questions they have. At these meetings, although all the suppliers have questions, most try to hold back anything that has the potential to reveal the direction of their response, so that the other bidders cannot guess what they will be proposing and try to outdo them. Thus, as mentioned above, some telcos have groups of people with great skill at asking the questions to which they need answers, and getting needed information without releasing too much to their competitors in the bid.

While the bid process is ongoing, it is also important that anyone involved with this process act very carefully in all interactions with potential vendors. Large telco equipment suppliers have been known to sue telcos on the basis that their people had interpreted interactions with telco employees as acceptance of their bid, whereas the telco people believed that they had never implied any decision. So when you are involved in one of these processes, ask questions and discuss possibilities for the equipment, networks, or services, but be very careful to assess how the vendor might interpret any discussion.

Another issue in the process is the issue of gifts from suppliers to buyers. Suppliers frequently provide benefits to the people who make buying decisions. These can range from lunch or drinks, to entertainment, to actual physical gifts. Most people are well aware that during the bidding process it is not acceptable to accept gifts from potential suppliers. However, the line beyond which something is deemed ongoing business versus a gift is not always clear. Perhaps one or more of the suppliers has an ongoing relationship with the buyer and, perhaps as a matter of course, the supplier, from time to time, takes the buyer out for an evening of dinner and entertainment, or business is often discussed over lunch, with the supplier paying for the meal. In fact, even during the bidding process, the supplier might discuss both information related to the open proposal and information related to the ongoing relationship at the same meal. Can the buyer accept the supplier's offer to pay for the meal? Many companies, and especially government departments, do not allow this, to prevent any perception of bribery. At least during the bid process, it is safest to avoid this, for the same reason. Clearly, physical gifts are not appropriate. But what about a pen with the name of the supplier? And if this is OK, where does it stop? The answer to these questions is not clear. The safest position is to refrain from any such giving during this period.

When bidding in response to an RFP, the supplier must also take great care. Of course, all of the requirements listed must be met because no noncompliant bids can be accepted, no matter how wonderful they seem. But besides just ensuring that every requirement is met, the supplier wants the buyer to select his proposal over all the others. Therefore, it must be written in such as way as to make the buyer aware of how well this proposal meets his needs and how well qualified the bidding company is as a supplier. Creating such positioning requires having a good understanding of the needs of the buyer, which usually requires some research into the position, the problems, and the environment of the buyer. This also requires industry research to understand what is available to the buyer from others so that the proposal can be positioned to show the advantages of the bidding company.

Generally, multiple bidders respond to an RFP, particularly when the RFP is issued by a company that purchases many products and services, which is usually the case for telecom companies, at least for those who own their networks. Occasionally, no bids are received, or none of the bids are compliant, which leaves the buyer in the position of having nothing to buy. If there are no bids, the buyer must contact some potential suppliers directly to find one that is capable of meeting his needs, and work out an arrangement. If there are bids, but these are not compliant, the RFP could be reissued, possibly with some changed requirements, in hopes that some compliant bids can be received. If the requirements are all necessary, perhaps the buyer can negotiate with some of the noncompliant suppliers to see if one of these could possibly build something to meet the requirements. Another option could be to reissue the initial request to a larger set of potential suppliers in hopes that one can be found with products that meet the requirements.

RFQ

An RFQ is a request for quote, and this is used when purchasing off-the-shelf items. It is subject to the same constraints and issues as the RFP.

VENDOR SELECTION

Once the responses to the RFP or RFQ have been received, selection of a supplier is the next step. This, of course, is completely dependent on the responses and how well these meet the requirements.

The first consideration is a determination of whether or not each submission meets the requirements of the request. In fact, if any submission does not meet all the requirements the buyer cannot accept that submission. And in addition to the technical specifications directly related to the product, service, or design itself, there will also be administrative requirements, such as the time by which the proposal must be submitted. Even breaking this requirement is not acceptable, and the buyer cannot accept any bid that is not fully compliant.

So the selection process involves an assessment of all compliant bids and matching these to the requirements. There should be some standard weighting assigned to the requirements, so that the assessment can be done fairly. Of course, the price generally plays a large part in the decision, but many other factors can be heavily weighted. The end result is not usually easily predicted.

During this process, there is often a need for some follow-up discussion between the buyer and one or more of the bidders, to clarify the proposal or obtain more information, or even to ask about possible new requirements. This negotiation process also needs to be handled in a manner that is fair to all vendors.

CONTRACT MANAGEMENT

Once the vendor has been selected and the details of the agreement discussed and negotiated, the real work begins. The buyer and the seller effectively have a contract, with the seller required to provide the goods and services in his bid, at the price and within the timing proposed. The buyer needs to manage this work, provide the seller with any information or access that is needed to allow him to complete his work, and give any feedback needed. Plus, of course, the buyer must pay for the work or material. In general, this is a process of contract management, including supervision, assessment and evaluation, cooperation, approval, and payment (or nonpayment if the requirements are not met by the vendor). This effort is the management of a contract. Some legal oversight will also be required to ensure that everyone is clear about the overall requirements, the portions that have been met at any given time or are still required, and the level to which the specifications have or have not been met. There could be interim payments that might be required at specific times, or on receipt of specific deliverables. All of this will take time effort, knowledge, planning, and careful monitoring. So all of this is clearly part of the overall project management.

CHAPTER 9

MANAGING RISK IN TELECOM PROJECTS

RISK MANAGEMENT

According to Wikipedia, "Risk management is a structured approach to managing uncertainty related to a threat, a sequence of human activities including: risk assessment, strategies, development to manage it, and mitigation of risk using managerial resources. The strategies include transferring the risk to another party, avoiding the risk, reducing the negative effect of the risk, and accepting some or all of the consequences of a particular risk."

For project management, the definition we will use is slightly different, in that the "threat" can be something with either a positive or a negative impact on the project.

Risk management is done in many organizations for many reasons, both within and outside of projects. In particular, organizations such as financial institutions are excellent at risk management because this obviously plays a big part in their business success. In this book, we refer to the risks thath might affect the management of these risks. Many of the techniques used for project risk management are the same as those used in general management or very similar to those, but they are used in project management to help to manage these problem situations within the project to ensure that they do not unduly affect the project ando derail it from the success track. Any knowledge gained in performing risk management on projects, in business or in life, will be useful to the project manager, and it is wise to take the time to stop and think about all possibilities before proceeding with the project.

The *PMBOK* gives six processes for risk management: risk management planning, risk identification, qualitative risk management, quantitative risk management, risk response planning, and risk monitoring and control. We will discuss the different activities that are all part of these processes.

The initial risk analysis should be done as early as possible in the project planning, because even a basic understanding of potential risks can be used to help design the project and its product in such a way as to minimize negative consequences to the project from foreseeable risks. The initial work will be used to help the team and the stakeholders to understand the risks and to allow the team to build the most workable preemptive responses to them. Some responses will be work that will be done beforehand to minimize the probability or the impact of a risk, and some will be work that will be done if and when a risk actually materializes. In either case, it is wise to have planned for this in the overall project work so that any adverse affect to the project is minimized, and also to ensure that there is enough time and money in the project plan to allow the risks to be effectively managed. Thus, it can be seen that there is a need for work to start early. However, the risk planning and management does not end there. This work will continue throughout the entire project. Although it is preferable that all the risks be identified early and plans be drawn up for these, it usually happens that some risks come to light as the project progresses, as the environment, the people, and the project evolve. Whenever a new risk is identified, impact assessment and mitigation planning must be completed for that risk, even if the project is well underway.

As in the previous chapters, all of this applies for all projects, no matter what their environment. And as before, the environment for telecom projects is very volatile: ever changing, with most directions not yet being clearly understood. All of this means that the level of risk for a communications project is higher than that of most projects before we even consider the particulars of the project itself. The project might be a very new concept for an established company that does not have the management mechanisms in place to deal with this sort of business, or it might be a typical project for a new company that is still shaky in making its way into the marketplace. In any case, the risks are higher than they are for projects in stable marketplaces. To add to this, many of the projects in telecom are based on technologies that are very new, often still under development. For these technologies, there is also high risk because the way in which they will operate or interoperate is often not known, and in some cases the technical standards within which they must work are often being set as the project progresses. Thus, there will be a large number of risks for the projects and some of these risks can easily be of major significance to the project. Risk management is thus even more important for telecom projects than for many others.

To compound this further, telecom projects are frequently multidisciplinary projects, with teams of people from many departments, with very different backgrounds and skills. This is necessary to build the full scope of a new product or service in a short time, but the degree of difference just adds one more set of risks to the project. Of course, the multidisciplinary nature of the projects occurs in most projects, not just telecom projects. So that in itself is not unusual in a project environment. However, as we have seen earlier in this book, the breadth of the background, knowledge, and focus of the people needed to create most of today's telecom services is considerably wider than it has been in the past. In the past, people from marketing, engineering, and various areas in operations, procurement, legal, sales were needed on a project team. Today, this list must also include people with artistic background or specific software skills in new areas that were never part of the project capabilities before. Ensuring that such teams can work well together, and can even understand each other, can be quite a challenge, and the risk is high that this will not go well.

Because of the complexities, it is important that the team sit down early and identify the risks that they will be facing. This is risk identification, and the identification of the risks should not be haphazard. However, even before this is done, the project manager might want to decide on a general strategy for dealing with the risks, so that once the risks are known, the strategy will automatically be applied. This is done in risk management planning.

The strategy should take a number of factors into account. First, what is the environment in which this specific project resides? If the company is a start-up with grand plans for a new technology, then the people involved are undoubtedly prepared to handle some level of risk. They have chosen to build a new company, which is a risk in itself, and because of this, they probably have a fair tolerance for operating in a risky environment. Whereas, if the project resides in a standard telco or another utility that is moving into a new telecommunications venture, there are probably many people in the project loop who do not have a tolerance for risk. The views of these stakeholders, which may also include the PM and the team, need to be respected. In a high-tolerance environment, it might be more acceptable to undertake more and larger risks, albeit with a good plan for how these will be handled, whereas in a more traditional environment, it would be wiser to eliminate or mitigate as many of the significant risks as possible, to avoid causing heartburn for those risk-averse stakeholders. The PM needs to assess the environment and make use of the information about the tolerances of the stakeholders in the planning for the risks.

Perhaps the risk-averse people will feel that some problems are more likely to occur or will have more disastrous consequences than the risk takers believe. Perhaps both groups will see the probabilities and impacts as being the same, but they will react differently to these. In one environment, more risks will be deemed to be significant than in the other, and probably more detailed plans and more aggressive handling mechanisms will be needed for those risks seen to be significant. Therefore, the risk management strategies will be different in the different environments. The PM needs to first assess the environment in which the project will occur, and then map out a strategy for dealing with the risks that will be appropriate for that environment. The lines will not be drawn as clearly as start-up versus established carrier. The specific stakeholders, the timing, the company strategy, the competition, and other such factors must all be taken into account in order to determine what strategy would be effective.

The strategy should first define how the risks will be classified overall—as significant or not, with as many levels of significance as would be appropriate for the project—and then clarify how the different levels of risk will be handled. Will there be a detailed contingency plan? Will there be an assessment of the costs of accepting versus the cost of doing something else? Will there be money and people ready to deal with the consequences if the risk should materialize? Will the risk essentially be ignored? Will people be trained to allow them to respond quickly should the issue arise, similiar, for example, to the handling of the risk of fire in an office?

It is good practice to initially determine the overall strategy before looking at the specific risks, so that the specifics do not unduly sway the risk strategy. But once the specifics are known, it is a good idea to map them into the strategy, then step back to verify that the strategy is indeed appropriate for the project. If there are significant discrepancies, a reassessment of the strategy might be in order.

The next step is to list as many as possible of the risks that the team will face in doing the project. This list should be prepared early for every project. The preparation of such a list can be seen as overhead, but having the list and using it to build the plans has the potential to save much more effort, energy, and heartache, not to mention money and time, than would be needed without the list. Some risks are not foreseeable, particularly in a project that is breaking new ground, but the experienced project team can usually anticipate most of the significant risks that will come up in the course of the project.

Risks can be identified by bringing the team together, maybe inviting some of the other stakeholders as well, and listing all the risks that each person can think of. This can be done by open brainstorming, or by using some sort of systematic approach. It can be done by the group together, or by having individuals submit inputs separately and then sharing the compiled version for reassessment by all. In a systematic approach, the PM might choose to provide some categories, asking people to think of risks within the categories. Such categories might include personnel, technology, regulatory requirements, project team structure, market dynamics, timing, familiarity with service model, stability of the environment, competition, and so on. Team members might add additional categories, and the company might have a standard set of categories that should always be considered. The end result will be a list that is as comprehensive as possible in identifying the risks for the project. If additional risks are identified as the project proceeds, they can be added to the initial list. But since the plans will be based on the early lists, these should be developed to be as comprehensive as possible. Creating such a list should also not take a lot of time if people are familiar with both the area of work in which the project falls and the company/environment in which the project takes place.

Qualification of the risks is an assessment of these risks from different perspectives. Will some of the risks be more visible than others, causing a need to watch these more closely? Is it important that some of the plans or actions for dealing with some identified risks be started in the near term, while others can be left till later? Perhaps a number of risks have a common cause, making it important to study the root cause of these risks in order to better minimize or prepare for handling them. The qualitative analysis should address all such aspects of the risks to allow the team to build the best plan for managing the risks to the project.

Once the team is satisfied that the list is as complete as possible, the quantification of the risks needs to be done. The end goal of this work should be the identification of those risks that fall into each of the defined levels of significance. These levels are generally not simplistic, so multiple factors will probably come into play in defining the criteria that will govern the categories into which the risks will fall.

The two main criteria for quantitative assessment are the probability that the risk will occur and the overall impact to the project if it does occur. Risks can impact the project in different ways, including impacting the cost negatively or (rarely!) positively, or the schedule, or the reputation of the project team or even the company. All of these need to be considered, and plans need to be created to handle the impacts.

In addition to the contingency plans and backup plans for individual risks, it is useful to also have some quantitative estimates regarding the overall risk probabilities and impacts, because these can be used to determine the amount of "contingency" in time or money that is required for the project. Contingency time should be included in the schedule, to allow the team enough time to be able to deal with those risks that will occur. Contingency money should be included in the budget to allow the team to cover the costs of the risks that will occur. The amounts of time and money needed will be determined using the information from the quantitative analysis of the probability of the risks and their impacts.

Since we will need to justify the inclusion of extra money in the budget and extra time in the schedule, we need to know and be able to explain the cost that the project would incur if each of the risks were to occur. That is the impact of the risk. If the impact is not one of time or money, but instead something less tangible such as loss of face or reputation, then some analysis is needed. Why is the loss of face a problem? Generally because it will cause the company to lose business, and that loss of business has a financial impact on the company. So the cost to the company of the occurrence of that risk is the amount of business that could be lost if the risk actually occurs. That is the amount that should be used as the cost impact of that risk. Similarly, the loss of face might have a cost in time, if the situation can be rectified or at least improved by having someone spend time working with the disappointed customer, or by some other mechanism. How much time might this take? That is the time impact that should be included for that risk.

Again, let us look at the telecom environment and telecom projects to consider why risk management is even more important here than it is in other projects, even though it is needed for all projects. In the initial chapters of the book, we looked at many things that are different today in telecom projects from the past.

1. As mentioned above in this chapter, the team is more multidisciplinary than was ever needed in the past. This in itself creates many risks that communication will not occur, that people will not be able to work with each other, that they will not understand each other, that the components of the project that they create may not interwork properly, and so on.

2. The environment in which the projects happen has completely changed. In the past, the environment was regulated, with perhaps some competition, but this was limited. The competition came from other companies who were known to the project company and who were in the same business. Today, there is little to no regulatory protection and considerable competition, and it comes from companies that are in many businesses other than telco. All of this creates risk that the customer needs might be misunderstood; that the requirements might be misunderstood so the product or service developed might not meet the needs, or might be too early or late; that someone else will better meet the customer needs, or offer something that customers like better; and so on.

3. This also creates the need for much more in-depth market research and understanding than that which telcos have traditionally needed. This creates a learning curve for the telcos that needs to be managed in parallel with the changing environment; and the unstable environment makes the research more difficult, bringing risks to the project in the area of performing and understanding market research.

4. The technologies are evolving at such a rate that no one company can keep up with even the knowledge of all the new technologies, let alone have time to implement them. The risk is high that the ones that might best meet the customer demands are not the ones chosen by a particular company, that the new technologies cannot be smoothly integrated with the existing ones, that the new technologies will not work well because they have been developed too quickly to have been tested to the levels that telcos have traditionally been used to, and so on.

5. The entire network architecture has also changed, requiring new learning by people who have been in the field, as well as by those now entering the business. More significantly, this requires very different back-office systems to allow billing, provisioning, data storage, ordering, testing, and maintenance. All of this must be developed in parallel with the changing environment, dramatically increasing the risk levels in the environment of the projects.

6. The types of services that the new technologies can offer are very different from the services that telcos have traditionally offered, and the number and variety of such services is increasing daily. This places everyone—traditional telcos as well as new entrants to the field—in a risky position as it is very difficult to know which services to build, how to design them, and how to get them working properly.

7. The business models for the services and products are also changing. The rates and the rate structures for the services, which had been relatively stable for many years, now change much more quickly. This requires not only market plans and sales packages, but also back-office system capability to be able to handle the new rates and rate structures.

This list is not exhaustive, but it certainly highlights the issues faced by everyone in the business of providing electronic communications services. Any one of these listed items poses significant risks to a project team, over and above the risks that are faced by teams operating in stable environments. In telecom, teams face all of these risks in parallel. Clearly, good risk management is crucial for every project in electronic communications.

For each risk, then, we first need to determine the impact both in terms of cost and of time. Next, we need to quantify the probability that each risk might occur. This is not an easy task. Generally, it is easier to start by thinking in less specific terms, such as whether the probability is high or low, possibly using a greater number of gradations. Most people feel more comfortable thinking in these terms. Once this has been done, with some thought and discussion, the team can assign some probability level to each gradation. Perhaps low risks will be quantified as 10% likely to occur, medium ones 40% likely, and high ones 70%. Note that for the high risks we would not use 100% likely or even something on the order of 90%. If the risk were 100% likely, then it is not an unknown at all; it is a fact, and it should be included in the plan as an activity. And even if it is "only" 90% likely to occur, this is close enough to certainty that it should be treated as a fact. So again, it should be included as a known occurrence, and the risk would be 10% that it would not occur.

The contingency to be added to the budget and schedule for a foreseen single risk is simply the impact of the risk on time or money, multiplied by the probability that the risk will occur. This is handled on a statistical basis, including all foreseen risks as shown in an example on page 93.

Companies frequently try to cut project budgets or schedules by removing some or all of the contingency due to a lack of understanding of what project contingency actually is. In calculating the contingency, the project manager attempts to clearly justify the need for any additions to budget and schedule, due to forceseeable risks, and it is highly probable that the contingency time and money will actually be spent. Contingency is not a hedge against the unknown (more about this later). The reason that contingency is often cut is that management often does not understand the reasoning behind the requested contingency figures, and believes that the team has already "padded" the budget or the schedule, so including more contingency is thought to be overkill. If they are correct that such padding has been included, then the contingency request is invalid. Therefore, the PM must ensure that the estimates included for the activity durations and for the activity/deliverable costs are as accurate as possible, and do not include contingency already. If this is done, then the contingency can be more accurately calculated and more easily managed.

The following technique can be used to statistically calculate the overall amount of money or time that should be included in the budget or schedule to deal with the foreseen risks of the project. Note that these amounts are calculated based on the actual risks to the project; a riskier project will need more contingency, whereas a more stable project should need less.

The technique for determining the amount of time to add to the schedule is the same as that for determining the amount of money to include in the budget. The example below shows the technique for calculating contingency money; the reader may use the same technique to determine time contingency. This is a statistical technique, which means first that we need to have a fairly large number of risks to work with; second, that no one risk can be an "outlier"; and third, that contingency figures calculated for a single risk in isolation are not meaningful in themselves.

One might ask how many risks are needed to make the statistics valid, but it is beyond this book to get into the concept of statistical significance. This should not be a problem though, because all projects are risky, and if the team takes care to list the risks they can predict as being possible, having too few risks to create a valid statistical analysis is very unlikely. So if the team compiles a thorough list, this technique should work very well.

An "outlier" risk is one for which the impact is considerably larger than that of all the other risks. If there are 300 risks in a project, with impacts ranging from $1000–$50,000, and there is one risk for which the impact is $150,000, then this last one is an outlier. If we use it in a statistical computation, it will skew the numbers. Also, if it does actually materialize, it could well absorb all the contingency funds for just this one risk. So if there is one risk that is an outlier, it is necessary to separate that one, and address it separately using some other technique.

The final point is that the individual numbers are not meaningful in themselves. This will be addressed by example after the technique has been described.

The method for determining the amount of money to include in the budget to cover the risks, known as the contingency, is as follows:

1. List all of the risks.
2. Quantify these. If you start by estimating the probabilities or the impacts in general terms such as Low, Medium, and High, then assign a numerical value to each gradation so that you will have a numerical probability and numerical dollar impact on the project for each risk in the list.
3. For each risk, multiply the probability times the impact to get a working dollar value for each risk.
4. Add the working dollar values. The total number is the number of dollars that you should add to the budget as contingency.

This is shown as follows:

Risk name	Probability	Impact ($)	Value ($)
Risk 1	0.1	1000	100
Risk 2	0.1	5500	550
Risk 3	0.1	10,000	1000
Risk 58	0.4	4000	1600
Risk 59	0.4	1000	400
Risk 60	0.4	8000	3200
Risk 410	0.7	2000	1400
Risk 411	0.7	1000	700
Risk 412	0.7	6000	4200
Totals		$260,000	$80,000

Thus, we can see that if we include no contingency money at all for this project, which has 412 potential risk items, we run the serious risk of going over budget, because some of these risks will occur, and for each one that does occur the budget will take a hit of at least $1000 and maybe as much as $10,000. In fact, the statistically most probable overall budget impact will be the $80,000 figure as calculated above.

So we do need to have some contingency money in the budget. If we are very risk averse, we would probably want to have the full $260,000 so that we would be covered no matter how many risks actually occur. No management will assign this much contingency to a project though, because it is clear that the probability that all of the 412 risks will occur, thus causing the team to need the full $260,000, is very low. Therefore, we need to ask for a number between zero and $260,000. Using the actual risks for the project, with estimates that are as accurate as possible for the probabilities and the impacts, we can find the number that will most likely be needed. In this case, that is $80,000. It is clear that the team will have to monitor the possibility of risk occurrence closely, and to use the money wisely, in order to ensure that they have enough for the actual cost of all the activities plus the cost of any risks that occur. Using the actual risks and probabilities in this manner demonstrates clearly why more contingency is needed for a more risky project, and less for a more stable one.

It is important not to get hung up and, more important, not to get senior management hung up, on the calculation of contingency for individual risks, as this leads inevitably to effort wasted in micromanagement. Let us look at risk 1. If this risk occurs, the team will need $1000 to cover the cost of handling the fallout. If it does not occur, the team needs nothing. Using this technique, we are including $100 in the budget for this risk. We will never need to separately manage the allocation of this $100 contingency. Although individual results may not be valid, the calculation of contingency over a large number of risks is likely to be statistically accurate, and should be added to the budget for contingency for the project overall.

CONTINGENCY IS EXPECTED TO BE SPENT

We can see that this money is in place to handle specific items—the risks on the list that we used to calculate the amount. There is not enough to cover all or even most of these risks if they should all occur. But if we have done a good job in estimating the likelihoods, then we should have enough to cover the ones that do occur. The money will be used but we need to ensure that it is used to handle these foreseen "known unknowns" only, and not for something else. Contingency should never be used for scope changes. If you decide to incorporate a scope change, you need to first find the new resources for it. If there is a requirement for more money, you must approach your sponsor, or your customer, or corporate management, or someone else for any resources you need. New money, new time, new offices, or new whatever will be required for the scope changes because they were not in your plan nor in your budget. They were not in your schedule at all. Scope changes cannot come from the contingency fund, as it is already spoken for as shown by the way it is calculated.

The risks are covered by the contingency. The name for funding that might be used for the scope changes is "management reserve." People, unfortunately, often use the two terms interchangeably, but these are actually two different things.

The contingency is to be used to cover the known unknowns, or risks, and only to cover these. It is part of the project budget, and part of the project planning process is to calculate the amount required, as shown above.

The management reserve is not part of the project budget. It is additional money that the PM must request, if needed, of course with a very strong business case and an explanation as to why it was not included in the first place. This is to cover unforeseen "unknown unknowns" that were not considered during the planning period, even if they really should have been anticipated. This money comes usually from the sponsor, or maybe from the customer. It can be used for change requests, since the funding for these is not part of the initial budget, but it will need to be added to incorporate the additional project scope if other committed deliverables are not to be left out.

As mentioned above, we calculate contingency for time in the same manner demonstrated here for calculating financial contingency. The time contingency must then be added to the critical path of the project, preferably in places at which the project is vulnerable, to help protect the schedule as much as possible.

With thorough planning, the team is ready to begin the project with full awareness of the potential risks, and a plan for dealing with these as needed.

However, as the work progresses, it is almost inevitable that more risks will arise. This could happen because the risk was by its nature unforeseeable, because the business environment in which the project is run has changed, or, most likely, because people simply neglected to think of some of the risks. Clearly, in today's electronic communications environment, things are changing around every project more quickly than people can anticipate the changes. This increases the overall risk level of the project and, therefore, the number of risks that must be handled. Once the contingency has been calculated and incorporated, it is usually difficult to build in more. This in itself is a risk, and may cause projects to fail even though they were otherwise well managed.

Risks that are defined later in the project are just as real as the ones identified early. Even if contingency money or time cannot be added, contingency plans can be put in place as the risks are identified, and the team can monitor the landscape to be ready for any of these which might occur.

Risk management is complex, and it takes thought, effort, and time. But that cost is usually far less than the cost to the project if the risks are not managed. In earlier chapters, we discussed many things that make projects risky in today's telecom environment. In assessing the risks for any specific project, all of the areas discussed earlier should be considered, to ensure that all possible risks are identified and managed.

CHAPTER 10

WHO TELLS WHAT TO WHOM?

COMMUNICATIONS MANAGEMENT

Excellent communication is possibly the most important aspect of project management. With good communication, project teams can manage quite effectively, maneuver through the many problems and issues that arise during project planning and implementation, and sometimes even find solutions for serious issues that might otherwise completely derail the project. With good communication, people can formulate the plans for the project, be kept aware of what is planned, and stay up to date on the status and the overall performance of the tasks can be coordinated. Communication is also a core factor in managing and motivating the team.

Communication is a key skill in projects as it is in life. Everyone communicates every day in many different ways. We learn communication throughout life, from parents, family, friends, teachers, and others we meet. Some communication skills are learned through formal training and education, whereas others come from experience. Some communication skills are culturally based, some are for interpersonal situations, and others are business related. The way in which each person communicates is to a large extent ingrained and very difficult to change when changes are needed.

Communication is key in developing teams and enabling the teams to work together in a unified manner. Teams thrive on a sense of togetherness, which is something that can only be built via communication. Projects, even more so than ongoing businesses, are based on effective teamwork and, therefore, effective communication is critical. Projects occur with significant time pressure, which tends to take the focus away from communications as people use their energy to find ways to complete the work on time. Communications are quite likely to be given a low priority under these conditions, in favor of getting the actual work done.

This is a mistake, because under such conditions it can be more important than ever to ensure that everyone is aware of what is happening and people are encouraged to continue their good work.

In telecom environments as in most others, departments such as sales, marketing, and public relations recognize the importance of communications. People who work in these departments tend to be people who enjoy communicating, and who have education or training in what to communicate, when, and how. They spend a lot of time, energy, and thought on their communications: what to communicate, when, how, by whom, to whom, and so on. On the other hand, people in engineering departments have highly sophisticated training in technical concepts, and they can communicate these very well among themselves. But technical people often do not have interest in communications about areas other than technical areas, and, in fact, they may even see such communications to be frivolous. They often do not have good insight into what information is appropriate for nontechnical audiences, and they may even feel that others need to know the technical details that are often not of interest to them. In telecom, most teams have at least a few technical people on the team, many consist completely of technical people, and often engineers are selected as project managers. This means that it is very important for project teams to carefully plan and define the required communications, to monitor these, and possibly to provide training in some aspects of communication to the team members.

In this chapter, we will look at communications from many different perspectives, all very relevant and necessary in projects. We will cover:

General communications
Project communications
Status reporting
Meetings
Motivation
Electronic tools for communications
Some suggestions

GENERAL COMMUNICATIONS

As mentioned, communication is a core component of good project management, and everyone communicates. Some people are better at some types of communication than others, and some people do not like to communicate at all. Some see taking time out from the "real work" to communicate as a waste of time. The project manager must deal with all of these people, and must ensure that all the relevant project communications happen on time, and at the right level, among the right people. For this, a good communications plan will be necessary, but it is also necessary to have a good understanding of communications and what makes communications effective. So we will first look at communications in general.

There are many methods of communication, including oral communication, written communication, nonverbal communication, and multimedia communication. These can be accomplished in many ways, such as in person, over the phone, via the Internet or an intranet, and other electronic means such as voice mail.

Everyone communicates and most believe that they are communicating effectively. However, it is often the case that the receiver of the communication does not see the same level of effectiveness as the sender. If we analyze communication, we might use a model that was developed by Shannon* for electronic communications. Shannon said that communication involves the transmission of information, or a concept from a source to a receiver. The information is encoded by the sender into a format that can be transmitted, and then sent through a medium to the receiver, who decodes the information. As the information flows through the medium, noise occurs, which can affect the quality of the information received. Shannon specified that it is important for the receiver to acknowledge receipt of the information so that the sender will know that the message has reached the destination, and he also recommended that the sender use redundancy in sending the message to ensure that the full, correct message did get to the destination. In Shannon's case, he was looking at converting the source information into something that could be communicated electronically, from one piece of equipment, such as a telephone or a dataset, to another at the receiving end. However, there is no reason that this model cannot apply equally well to human communications. Then, if there happens to be a network and/or equipment in the equation, this just adds one more complicating factor.

With human communications, the originator of the communication has information or an idea that he or she wants to convey to one or more others. This person converts this information into words and collections of words, maybe also pictures or other sounds for a multimedia communication, and "sends" this information to the other(s). If the communication is oral, the sending could involve meeting in an office, meeting room, or restaurant. In this case, the medium through which the message will be transmitted is air. It could be that this information will be sent through a telephone, or a computer, in some electronic format, in which case the medium will be a collection of air, equipment, and transmission facilities. The receiver, when he hears the words and sounds, will form an understanding of the information transmitted. The conversion used by the sender to translate the information or concept into words, sounds, or pictures is based on a model that the sender has learned is the right one to ensure that the receiver will understand the concepts in the same way that the sender understands them. The conversion model used by the receiving person is based heavily on the experience of the receiving person, and this experience must necessarily be different from that of the sender. Therefore, the two conversion models will never be exactly be the same. Maybe it is a wonder that

*Shannon, C. E. A, 1948, Mathematical Theory of Communication, *Bell System Technical Journal,* Vol. 27, pp. 379–423, 623–656.

any communication between even two people, let alone among groups, actually conveys the information that it was intended to convey. The closer that the two parties are in experience, including language, culture, job experience, life experience, and current life focus, to name some of the variables at play, the more closely it is that the received message can match the one sent. This is without considering the impact of any noise in the transmission. The noise can be actual noise within the room in which the conversation or presentation takes place, as well as other "information" that enters the heads of the two parties. Other information may come from visual stimulation that is occurring within sight of one or both parties, sounds that are in the background, as well as thoughts in the heads of the two parties. Any noise has the potential to impact the information that is being transmitted. Therefore, the redundancy suggestion is a good one. When giving a presentation, a form of redundancy is already there, with the verbal information synchronizing with the slides. Even a conversation or phone call could be followed with written information to ensure that the receiver does receive the full information as it was intended. Acknowledgement is also a very good idea. Many times, people say something or send a letter, note, or report without getting an acknowledgment that at least something has been received. With all the different possible ways in which the communication could be waylaid or muddled, it is a very good idea to ensure that an understandable and accurate message was received.

If a message is not received, the sender has the responsibility to ensure that the information does get to the receiver. The receiver may not even be aware that something is being sent, so it is necessary for the sender to assume this responsibility. However, if the message received is unclear to the receiver, the sender will not know this unless the receiver identifies this fact. Shannon does recommend that the receiver provide feedback, which can assist the sender in understanding either that the receiver is not clearly understanding the message or, worse, that the receiver has an understanding of the information but it is not the message that the sender was trying to send. If we take a few minutes with important communications to ensure that the right message was received, we can avoid many problems on projects.

Let us add one more extremely important component to the communications equation. This is listening. Listening seems to be a skill that is very difficult to develop, even for people who really enjoy interacting with others and find the others to be very interesting. How can a message be fully communicated effectively if the receiver is not truly listening? Project managers could do well to learn to listen themselves, giving them strong information about what the team members, the customer, and the management expect, as well as an understanding of the status, the issues, and their potential solutions, and the available support. Project managers can also help enhance the project experience by encouraging others to also listen, and finding ways to set up the environment so that it is easy for people to do so. People need to ensure that they listen with the intent of gaining understanding. The focus cannot be on listening in order to shoot the speaker down or push back with an answer. There must be a focus on really hearing and understanding what the message is. Only then can people work together to find true solutions for the good of the project, the company, and the team.

Communication is needed within projects for many reasons. The information about the project itself must be captured and conveyed, first from the project initiators to management, then from the sponsor to the project manager, and later from the project manager to the team. The information, in some form, must go to the functional managers to justify the need for using their people on the project team, to the key stakeholders, and to any potential suppliers to allow them to understand the requirements for them. Each of these communications has different requirements and, in many cases, quite different information will be conveyed. Given the nature of communications as described above, it is necessary for the sender in each case to consider who will be receiving the information, and to determine what information this person needs to receive. The information also needs to be configured in such a way that it will get the attention of the receiver, and that it will be easily understood by the receiver.

This means that the sender should learn something about the receiver, in order to design the right communication. Information going to senior management needs to be crisp, clear, and to the point. The details should be either not included, or relegated to appendices that can be read if the receiver desires that information, but otherwise left out. Information going to team members, though, must be very detailed in order to ensure that each one has the full clear understanding of the project and the work that needs to be done. Within a project, there is a need for different communication packages to be used for different people and different purposes. Some such communication packages are:

- Project charter
- Project scope description
- WBS
- Logic diagram showing the work flow
- Gantt chart showing the schedule
- RFI and/or RFP
- Supplier bids and proposals
- Requests for funding
- Requests for personnel
- Status reports
- Contracts
- Minutes of meetings
- Customer presentations
- Invitations to meetings and agendas
- Product descriptions

- Product manuals of various types
- Project budget
- Project funding requests
- Project account tracking reports

Each communication serves a specific purpose, and each goes to different people. Some follow predefined formats, and some must be designed as they are created. It is clear that there is a need for a lot of thought, effort, and planning to ensure that proper communications happens.

Another reason for communicating with the team and the stakeholders is to initially set their expectations for the product and the work, and then to manage these expectations as the project proceeds. This can become quite difficult if the project does not flow as expected, so some good communication skills will be required within the team.

This can be extended as well into building a team spirit and a sense of belonging to the project team. Teams that work together can do so because they understand each other and have developed techniques of working together so that things go more smoothly. Communication is an important tool in building this rapport.

Studies show that in addition to the verbal information we send to one another, we send additional information that is nonverbal. In fact, a fairly significant percentage of the communication that occurs comes in nonverbal form. The way a person dresses, walks, stands, or postures each convey information about the person. What is acceptable or commands respect in one environment is not accepted in another. This makes it important for the team members to understand their audiences and to think about the information that they will convey via means other than their words. Words said with a smile can elicit one response, whereas the same words said forcefully with a stern look will carry a very different message.

MANAGING THE COSTS

The information above provides the project manager with quite a lot of data for use in managing the costs of the project. Creating the detailed plans for the cost is the first step in managing the costs. It is necessary to understand how much money has been allocated for each of the deliverables and activities in order to be able to understand when the spending is getting out of line. Both the budget and the planned value should be used, and this information should be shared with the team members so that all will be aware of the restrictions under which they should operate. However, just understanding the information does not constitute controlling the costs. The team should establish a cost management plan which identifies the limits at which the costs will be deemed to be starting to get out of hand, and which identifies what corrective action should be taken, and by whom, if this problem should arise. Also, a member of the project team needs to be tagged to monitor the costs, probably at a deliverable or activity level as well as the bottom line. This person should raise a flag if things seem to be heading in a bad direction. The EV can show this at the overall project level, but some monitoring at a more granular level can help the team to be prepared and to act quickly to catch problems before the situation becomes too bad.

With all of this in place, the project will have the best possible chance to succeed in meeting the cost requirements.

PROJECT COMMUNICATIONS PLANNING AND THE COMMUNICATIONS MATRIX

As discussed, a good communications plan is very important for a project. For every project, there is a need to define who will be responsible for providing the different communications, and this includes both preparing the material and giving the report or presentation, or otherwise conveying the message. For every communication that needs to occur, there is a need to understand who is the receiver, who is the sender, what medium should be used, and what mechanism or format is appropriate, in addition to planning the content.

Stakeholders need to understand many things about the project including: why this project is important, what the nature is of the project, who is involved, what the current status and issues are, and what information needs to be known, by whom and when. Good communication will allow others to help. This goes beyond the team members to include people such as suppliers or the project sponsor. Good communication will allow people to be aware if and when they are needed to help and also to allow future teams to avoid similar problems.

In order to ensure that all of this information will be in the right places at the right times in the right format and with the right content, every project should have a communications plan. Some of the communications will be related to the product, such as product specifications or manuals. Some will be related to project status, such as weekly status reports or funding/spending reports. Some will be related to how to do the work, such as written procedures. All need to be planned.

For every communication, the team needs to understand who will initiate or prepare it, who will receive it, what information or content is needed, what format should be used, what medium is best, and when it should be sent or received. This plan can be prepared as text, but some teams might prefer to use something that is a little easier to work with, such as a communications matrix (Figure 10-1).

The columns of the matrix can be used to list one aspect, such as communication initiators, and the rows can show another aspect, such as the specific communications. Then in the boxes, all the relevant information can be shown either in detail or via codes. In the sample shown in Figure 10-1, which is only partially completed since it is a sample, some of the codes are:

W—written by this person

A—approved by this person

	Joe B, Sponsor	Sandra, PM	Robert	Endie	Chris	Lila
Prepare Charter	W, June 10	A. June 29, All team members, their FM's, P	I, July 15	I. July 15	I, July 15	
Agenda for kickoff meeting	A, July 10	W, A, July 10, team, P	C, July 13	C, July 13	C, July 13	
Minutes of kickoff meeting	I, July 23	A, July 20	C, July 19	W, July 16, team, Joe	C, July 19	
Minutes of brain-storming re product	I, July 28	A, July 24	C, July 23	W, July 19	C, July 23	
Minutes of scope planning meeting		A	W, July 27, team, FM's	W, July 27	C	C
Scope statement	I	A	C	W, July 30, team, Joe, FM's, customer	C	C
Minutes of data-sharing meeting	I		C	C	W, Aug 5, team, P	C
Work breakdown structure	I	A	W	C	C	C
WBS assumptions	I	A	C	W	C	C
Customer presentation re product	I	A	C		W	
Management meeting report	A	A	C	C	W,	C
Management meeting presentation	A	A, Present	C	C	W	C

Figure 10-1. Communications matrix.

C—gives consent

I—receives this as information

P—post on team intranet site

The dates show the date at which the information must be available, and in the box for the person who will issue the information the recipients are listed. In this way, a large amount of detail can be conveyed very efficiently. It is also good practice to specify the format for any communication for which this might not be obvious. In large organizations that typically operate on a project basis, there will usually be templates available for all commonly used documents and other communications vehicles. In many cases, the project might have literally hundreds of communications, and with large teams the number of columns would also increase. But it is easy, using the matrix, for each person to determine his or her communications responsibilities and it is also easy for the PM to monitor the activities for completion.

STATUS REPORTING

Given the nature of projects, status reporting is one of the most common communications that occurs. Every project needs status reporting. Probably every team member will have to prepare a report or portion of a report at some time during the project, and most will submit reports at regular intervals during the project. The PM needs to clearly define what is required and when, and this reporting should be minimized to ensure that all information prepared is really needed.

Reports submitted to the PM to keep the PM informed of the progress need to include enough detail to allow the PM to understand the status of the work and to be aware of any issues that might need to be resolved. However, if the PM is closely involved with the work, he is probably fairly aware of the overall status of the elements, and would not need to have a lot of detailed documentation for his own information, but that is usually not the case for all team members. In most project organizations, these periodic reports, or summaries of them, are distributed to the team. The reports are often archived for use in planning future projects. The level of detail will vary: perhaps a checklist could suffice in some areas, or a percent-complete box (more on this later). The PM will also need to prepare reports for management, and possibly reports for the customer from time to time. Rather than having these reports prepared from scratch, it is more efficient (and more accurate) to reuse as much information as possible from other existing reports. Therefore, it is wise to plan for this, considering the format and the content that should be used for the management and customer reports, and structuring the internal reports in such a way that contents can easily be copied or summarized for preparation of the external reports.

Collective inputs to a central repository can be quite effective, as they have the advantage of being immediately available to anyone with authorized access to the data, and being readily available for the preparation of summary reports. In this way, the team, and perhaps some key stakeholders, can always be aware of the progress of items that precede their own work or which are critical for their operation. This allows the person doing the work to report only once, without having to answer multiple questions and provide information in different formats. This data can be used to build reports that go beyond any one area. If the repository is set up with the capability to summarize or manipulate data in addition to capturing it, this can save further time on all parts for the preparation of reports in different formats, for different purposes.

There is always a danger in reporting the percentage of work that has been completed. Unless there is a known and accepted measure that defines the percent completed, it is left to the individual doing the work to make this estimate. People have different approaches to making estimates, and different views of the relative amount of work that has already been done. An optimist might overestimate the amount of work that has already been completed. People in general tend to overestimate the amount of work completed until the end approaches, when the percent complete growth tends to slow dramatically as the last stages of work are done and they actually consider what is required in the remaining work. The rare pessimist might err the other way. Too subjective a measure of percent completion yields numbers that are not consistent from one project area to another. It is best when this technique is used to have people estimate the amount of work still to be done, rather than the percent complete. Not only does this force people to more clearly consider details of the remaining work, but it will also help in pulling the estimates closer to accurate values.

MEETINGS

Much has been written about running effective meetings, and there are many things that should be considered to ensure that meetings are of value to everyone involved. However, the for the most part this has not been shared with people in business and people running projects. It is rare that meeting are as effective as they could be, because many people still seem to consider meetings to be a waste of time, an unpleasant overhead, and an interference with their ability to get work done. At the other extreme, and worse still, is the thought held by some people that holding a meeting is an end in itself, rather than a means to get things done.

Meetings need to be used as an effective tool in enabling people to complete the project. They need to be used to keep people informed; to allow for discussion of plans, status, and issues; to generate solutions to project related problems; and to allow team members to understand each other and build the rapport that will enable them to work effectively together, but this needs to be done in a way that uses everyone's time effectively and produces results of value to all team members as efficiently as possible.

The key to achieving this is good meeting planning. Good meeting planning involves a number of steps, and generally also involves more than one person, to ensure that all the aspects are included properly. The planners must ensure that the agenda includes all items that need to be considered by those attending, so that people can leave with everything they need to further the work; that the material is presented in the right order, to prevent repetition; that the right people attend the meeting so that nothing is missed and decisions can be made; that everyone who will attend the meeting is aware of what is required of him or her at that meeting; that the time plans for the agenda items are respected so people can accurately understand when they can return to their work; and that the minutes of the meeting are prepared, with the right information at the right level of detail so that they will be useful. To ensure all of this, the preparation for the meeting will usually take more time than the meeting itself, but the preparation need not involve all of the attendees, and each person will prepare his own portion with minimal involvement from others. This makes more effective use of everyone's time. Let us address each of these points individually.

Preparing the Agenda

The agenda for any meeting is much more than just a list of topics to be discussed. The agenda is the guide that tells the people attending the meeting what will be discussed, when, by whom, and why. The topics for discussion are obviously at the core of this, but even these must be selected carefully. In order to ensure that the right material will be discussed, it is first important to understand the purpose of the meeting. The meeting agenda, therefore, should start by identifying the meeting objective, to ensure that all involved will understand it and can make their own decisions about whether or not their attendance is important. If there is no good reason to hold a meeting, then doing so will clearly waste the time of the attendees and give meetings the bad name that exists today in the minds of many. The person who chairs the meeting will decide on the meeting objective, but it is often useful to involve one or two others in building the agenda, to ensure that others are also aligned with the need and the requirements.

Below we will discuss the roles of the different people involved in making the meeting run smoothly. One of these people can be a facilitator. If the meeting has a facilitator, this person should work with the meeting chair, who is often the project manager, to define the purpose and build the agenda. When the objective is understood, the agenda topics can then be selected. As each is selected, it is important to consider why this topic is to be included, how it should be handled, who needs to be involved, and what outcome is expected at the end of the time devoted to the topic. Then, when the agenda is drawn up, as each item is identified, the method in which the item will be handled can be identified, along with the time for the item, the name of the person who will handle the item, and the expected outcome. Armed with this information, people can be better aware of what to prepare for the meeting, what they need to bring, and what will be expected of them at any time. People who are not needed for some items may chose to leave the meeting during those items, so as not to waste their time, and this expectation needs to be shared by the meeting organizers and attendees. Because the time required for each item will be identified, and if the meeting organizers ensure that the times are adhered to, people will be able to judge the amount of time that they must spend at the meeting beforehand. In addition to determining the nature of the items important for the meeting, the planners must also determine the order in which they should occur. If information on one topic is needed in order to effectively understand or make decisions on a subsequent item, then the first topic needs to be covered first so that the second one can be handled effectively. Someone needs to sit down and think through the details of the items in order to determine the best flow.

In addition, if there is background material that the attendees should have read before the meeting, this material should be sent out with the actual agenda, with instructions as to what it is and why it should be read before arrival at the meeting. Then everyone can arrive at the meeting prepared with this knowledge, rather than expecting to have this information presented during the meeting. A sample agenda is shown in Figure 10.2, to illustrate these points.

Inviting the Right People

If presentations are to be made, information is to be discussed, or decisions to be made, then it is important that the right people be at the meeting. Those people who will make the presentations, those who need to hear the presentations, and those who must be involved in decision making, either because they have relevant information or they need to agree to the decision, must be present. If the right people are not there, the item cannot be fully completed, thus making inefficient use of the time of those who attend the meeting.

The meeting planners need to determine who is needed to present, discuss, or be informed about each item, and they need to ensure that those people are invited to the meeting, or at least to that part of the meeting during which their presence is needed. The invitees should, of course, ensure that the planners are informed about whether or not they can attend. The planners should check ahead of time with all critical attendees to ensure that the meeting schedule is acceptable to them. The planners should call the meeting only for a time at which all key players can attend.

Informing People of Their Roles at the Meeting

If people are to be expected to participate in a meeting, they need to be made aware of what they are expected to do and why they have been invited. Some people will simply be attendees, who will listen and participate in discussion or decision making. Some will be requested to report on specific things, and others will be expected to take notes, keep the meeting on track, ensure that the time allocations are met, and so on. If so, these people need to be made aware before the meeting of the expectations so that they can come prepared and be present during any portion of the meeting for which they are needed.

Everyone should respect the time of all other attendees by ensuring that they arrive on time, come prepared, bring any required information or material, help keep the discussion focused, and keep any commitments they have made regarding the meeting. When everyone does these things, everyone gains in productivity and in satisfaction with the meeting. No one should be expected to attend a meeting at which he or she is not needed, and people should be excused from portions of meet-

Agenda				
Project M Advisory Board Kickoff Meeting				
Wednesday, 17 June 2009 9:00 am–3:00 pm				
Meeting Chair: Celia Desmond				
Facilitator: Carole Summer				
Notetaker: James Ballantyne				
Scribe: Susan Thor				
Timekeeper: Sandy Miller				
Invitees: Allan Green—VP Technologies German Bell—Director, Customer Relations Ashley James—General Manager, Sales Ted Mumbai—General Manager, Broadband Frank Reeves—VP Corporate Communications				
Project Team: Celia Desmond; Ronnie Felt; Susan Thor; James Ballantyne; Janet Corn; Sandy Miller, David Stickney, Arjan Taparia, Dennis Bean, Rolf Walcott, John Norblett				

Start Time	**Subject**	**Presenter**	**Methodology**	**Objective/ Expected Outcome**
Wednesday				
9:00 AM	Welcome, introductions	Ronnie Felt	Conversation	Introductions
9:10 AM	Background: Project M drivers	Celia Desmond	Presentation	Create awareness of project background
9:45 AM	Checking the competition	Carole Summer	Presentation	Understanding of industry direction
10:30 AM	—Break—	—	—	—
10:45 AM	Development process for Project M business case	Susan Thor	Presentation	Understanding of detailed steps in developing the business case for the project; selection of key parameters for objectives *(continues)*

Figure 10-2. Sample agenda.

Start Time	Subject	Presenter	Methodology	Objective/ Expected Outcome
Wednesday				
11:15 AM	Current plans for future of Project M	Celia Desmond	Presentation a nd discussion	Understanding of Project M scope, benefits, and future directions; decisions on 3 potential direction issues
11:45 AM	Promoting Project M products	James Ballantyne	Informal presentation with sample marketing materials	Understanding of marketing direction and industry response
12:15 PM	—Break—	—	—	—
12:30 PM	—Lunch—	—	—	—
1:40 PM	Advisory Board member comments	Celia Desmond/ Ronnie Felt	Round Table	Board recommendations understood and agreement reached
2:20 PM	Advisory Board action planning	Ronnie Felt	Conversation	Determine and assign individual next steps
2:40 PM	Session wrap-up	Rolf Walcott/ John Norblett	Conversation	Set potential meeting and/or conference call dates
2:50	Benefits and cons	Celia Desmond	Input from each Advisory Board Member about value of this meeting	Better prepared to run a more effective meeting next time
3:00 PM	Meeting close	—	—	—

Figure 10-2. *Continued.*

ings if their presence is not required. However, this can only be done if everyone re-spects the time schedule and ensures that the attendance and the agenda items occur as needed.

Using the Meeting Time Effectively

As shown in the sample agenda in Figure 10-2, running an effective meeting is best done with the help of a small number of people, rather than placing the entire burden on the person chairing the meeting. Each has a specific role to play, and if these roles are handled properly, meetings can run quickly and quite efficiently. Not every meeting needs to have someone in each of these roles, and for some meetings it would be practical that one person could take on more than one role. But the responsibilities for each role differ, so it is important to understand what is required in each.

The first role is the meeting chair. Generally, meetings are chaired by the PM, especially if they are meetings of the full team or all team leads. The function of the meeting chair is to lead and control the meeting. The chair decides whether the meeting is to be informal, as most meetings for telco projects are, or formal, perhaps following some rules such as Robert's Rules, which are far too structured for most telco meetings. In either case, the role of the chair calls for the chair of the meeting to remain unbiased in discussions and to get all points of view on the table. The chair should not advocate one view over another. When there are items that the chair does want to speak about, he should relinquish the role of chair for the duration of the discussion, asking someone else to fill in for him in that role to allow him to speak about the issue on the table. The meeting chair must also supervise the meeting planning and organization, ensure that the right people have been invited, ensure that all meeting roles are filled, and ensure that everyone has a voice at the meeting. The chair also ensures that the agenda is followed as it is written, or that the attendees approve changes to the agenda, even if the changes are proposed mid-meeting.

The second role is that of facilitator. It is not necessary to have someone fill this role, but many PMs find it handy to have someone to work with them for meetings that are critical, complex, or address complex issues. The facilitator essentially supports the meeting chair. The facilitator is usually someone from outside the core team, and definitely someone who is seen by the team as being neutral on team issues. The facilitator helps to plan the meeting, asks the chair many questions, and tries to anticipate the difficult questions that might arise at the meeting so that any needed planning for dealing with these can be put in place prior to the meeting itself. This person asks questions at the meeting in order to ensure that all the information is on the table. The facilitator can be positioned to ask questions about information that should be brought forward when the PM or chair knows that no one is likely to bring this up at a meeting unless asked and the chair does not want to ask the question himself. If the chair wants to be able to hand over the role of chair during the meeting in order to participate in discussions, the facilitator is a likely choice as his temporary replacement.

The next role is that of notetaker. This is the person who takes the minutes. In order to ensure that all the key information is captured properly, this person should be someone who is otherwise not expected to have a very active role in the meeting. If this person plans to give a presentation or be very active in a discussion, he might ask another attendee to keep some additional notes during that part of the meeting to ensure that nothing is lost during the time that his attention is distracted from noting the important aspects of the meeting.

Then there is the scribe. This position is different from that of notetaker, because the scribe works on a padboard at the front of the room, or today possibly on a computer with the screen projected so that the attendees can see it. The scribe notes key ideas in point form, giving the group a visible, real-time record of the flow of the discussion, the ideas presented, and, possibly, actions assigned or decisions made. The notes written by the scribe are often used by the notetaker as supplementary information in the preparation of the meeting minutes.

The last organizational position is that of timekeeper. This person should have a clock or at least a good visible watch. Each agenda item is assigned a start and end time. The timekeeper monitors the pulse of the meeting information flow, watching to ensure that at the current rate of development, the item will complete on time. If things appear to be going off track, the timekeeper's role is to identify this fact, in hopes that the person leading the item, perhaps with assistance from the facilitator or chair, can bring the item back onto the time track. In cases in which the time limit cannot be met, the timekeeper must ask the group to make a decision on how to handle this. Options can include continuing with the item at the current rate, extending the meeting time by the amount of time the item will run over. This should be done only if those who need to be available for the last items can manage this. Otherwise, the team could agree to stop this item at the designated completion time and deal with it later either offline or at a future meeting. If this is to happen, the person responsible for the item should ensure that at least the information and decisions that will be needed to handle the remainder of the meeting items are completed, so as not to impact other items at that meeting. Another possibility is to continue with the item to completion, making up time later by cutting short items that fall later in the agenda. Some combination of these solutions could also be used. It is important that the attendees discuss the potential solutions, and agree on the best one. This discussion in itself will take time, so it is always preferable to have everyone act in a manner that will allow every item to complete in the time allocated.

The responsibilities of the attendees are to confirm attendance, to appear for the meeting or give enough advance notice of absence that the meeting could be rescheduled if necessary, to come to the meeting prepared, to listen to presentations, and participate in discussions and decision making. Perhaps the most important responsibility, which is inherent in all of the above, is to respect the other attendees and their time.

When all of this is done, meetings can be enjoyable, covering all the required material and producing all the required results while making the most effective possible use of everyone's time.

Preparing the Minutes

Every meeting should be documented in order to have a record of the results. At one point in time, minutes often consisted of information about everything that was said at the meeting and who said it. Writing this sort of long document is not generally useful in the telecom environment and, in fact, doing so can be negative. The potential for error is high with the high volume of writing and the level of detail, and the people receiving the minutes are likely to not even read them because they are so long. It is far better to document facts such as the outcome of key decisions, notes about any commitments made, and identification of any action items assigned with the name of the person assigned and the date it is due.

In the action items, be very specific. Document clearly what is to be done, by whom, and by when. This is one place where additional information might be useful. The minutes must be clear to all so that the right work can be done, on time, and so that everyone is aware of who will be doing the work. As a meeting follow-up, it is also wise to double check that everyone who has been assigned action items agrees to complete these as required. This also holds true for decisions. Every decision should be very clearly documented and highlighted so that everyone is aware of them.

At the end of every meeting, time should be spent discussing how the meeting went. People should refer back to the meeting objective and decide whether or not it was met. What else was accomplished and what was not accomplished that should have been? How could the running of the meeting be improved in order to make better use of people's time at the next meeting and to ensure better outcomes in the future? The learning from the running of each meeting should be shared and taken into consideration for the planning of future meetings. It does not take much time to do this at the end of a meeting, but the savings at future meetings can be significant.

Minutes can be stored in a central online repository, if there is one, to allow anyone with access to the database to read them. This saves the team from sending the minutes to everyone who needs them, although, in many cases, it is wise to also send them out, to trigger those very busy people to at least be aware that they are available.

MOTIVATING PEOPLE TO COMMUNICATE PROPERLY

In projects, the team is usually a temporary one, with each person working on something that is not his or her usual job. There can be many pressures on the team members, such as working for multiple bosses in some organizations, or a desire to shine on the project, or a desire to return quickly to the usual job functions. The project manager sets expectations that are often different from those of the usual supervisor, and the project manager needs the project work done as per project specs, within the project timelines and within the project budget. Although some team members might find this motivating in itself, others can find the situation very stressful. In order to meet the project goals, team members must be motivated to do the project work. The job of motivation falls to the project manager, and this cannot be done without good communication. It is important that the PM communicate clearly what his expectations are in terms of scope of work, timelines, budget, quality of product, and so on. It is important also that the PM clearly communicate feedback on the work, with as much positive feedback as possible in order to motivate the people to continue to improve and produce good results. The PM can also help to motivate good work by communicating the praise not only to the employee, but also to the employee's regular supervisor, so that others who have influence over the employee can be aware.

Project managers can also motivate people by involving them in decision making or planning for areas that impact them. This might involve brainstorming or serious discussion of the project deliverables, options, and work. The PM can encourage innovation and good ideas by employing one key technique in communications—listening. This needs to be followed by feedback that both shows that the PM heard what the team member said and gives credence to the proposal being put forward. The PM needs to work with the team member to find ways in which some of the proposals the team member puts forward can be used to further the project work.

ELECTRONIC TOOLS FOR COMMUNICATIONS

In any area today, people are either already well aware of the benefits of using electronic tools for communications or they are rapidly learning these benefits. In the telecom industry, people are particularly tuned to the latest available technologies. This is not to say that everyone in telecom uses, or even likes to use, the latest electronic tools; in telecom, as elsewhere, there are early adopters, people who go with the flow, and people who prefer to hold on to the older tools that they know well. Almost everyone in the industry is at least aware of the types of tools that can be available in support of projects in the telecom environment.

The tools available are many, ranging from the standard office-type programs for creating word-processing documents, diagrams, presentations, and spreadsheets, to applications that are strictly project oriented, such as Microsoft Project or MAC Project. In addition, electronic storage and retrieval systems are very useful in storing project information that can be made available to those with approval to view different packages of information. Online archiving of project files is only one example. Projects carried out by a team in multiple locations benefit hugely from communications by e-mail and instant messaging. In most companies, online meetings using Internet meeting applications outnumber the ones that require everyone to be in the same room. High-quality telepresence facilities provide lifelike audio and video links for companies with deep pockets. The potential today for saving time and making work more efficient via the use of computer and communications technologies is almost endless.

One issue that teams might face in using rapidly evolving communications tools is lack of compatibility between systems or applications. This will tend to be an issue only in cases where project teams span multiple companies, a rather rare occurrence. Any successful company in the telecom sector can be expected to have an IT department whose job is to ensure compatibility and ease of interworking among the supported systems and applications.

The effective use of up-to-date electronic tools for all aspects of project communications is to be expected in the telecom industry.

MONITOR AND CONTROL

Every project has a work breakdown structure that identifies deliverables and all the activities needed to produce these deliverables. The work is assigned to specific individuals, to be completed according to well-defined and usually quite challenging schedules and within the required budget. It is important to all team members that every work package be completed properly, on time and within the assigned budget. The accountability for ensuring that this happens lies with the PM. The first step in this process is defining the work. This should be done by the full team or the relevant people from the team for each area, with some degree of advice or direction from the project manager. This in itself requires communication, to ensure that the right deliverables and work packages are included. Next, the PM needs to work with the people who will actually perform the work to determine the time required. This again requires very clear communication, as only the person who will do the work really understands how long he or she will need. The PM needs to ensure that the time estimates are the estimates for doing the actual work, without any contingency included, so that the PM can control where the contingency time is assigned. The fact that contingency time is included, and the locations of this time, also need to be communicated so that the team members and the PM all have a full understanding of the true situation. The PM then needs to get commitment from people that they will do the work required in the desired manner, and then to ensure that they have all the tools, abilities, and conducive environments to allow them to proceed with the work.

Once all of this has been done, understanding should exist. It then falls to the PM to monitor and report the status of the work, which again requires communication. The PM must ensure that everyone understands what information he needs to receive, when, and in what format. This status information will be compiled by the PM to get a full view of the project overall.

When planning for quality management and outputs, the PM should set control limits beyond which action to take to get things back on track. When these limits are reached, the PM must also communicate this fact, and ensure that everyone affected understands the change in work that is now required to help get things back on track.

SOME SUGGESTIONS

We communicate in order to transmit information, receive information, use the information to solve problems, influence events, or create an image in the minds of the listeners. Communications are often in place to help someone to influence another person or some event. Therefore, it is very important that the right communications happen at the right time, and that they accomplish what they are intended for.

Effective communications can be the mechanism to ensure project success; poor communication can cause the failure of the project and the downfall of the team. Everyone on the team should be aware of this fact and encouraged to use the most effective communications techniques and technologies whenever possible. This section puts forward a few tips for making communications effective and clear.

To ensure effective communication, planning is critical. In this chapter, we have discussed numerous different aspects that must be thought through, with conscious plans made to provide everyone involved with correct and relevant information about what is required, when, by whom, and for whom, in what format and medium. Ensure that everyone involved participates in the planning, is aware of the plans, and commits to making them happen.

When communicating, both in person and in writing, remember one of the most critical aspects of communications, one which many often forget: listen! Listen to the customer requirements; listen to the sponsor's vision; listen to the team members' plans and concerns, listen to suppliers when they explain their concerns, ideas, and issues; and listen to the functional managers regarding their problems with releasing their people. In short, truly listen to everyone, and try to find ways to ease their problems so that they can better support the project. Try to determine not only what their words convey, but what might be behind the words. Not everyone speaks clearly and brings the important issues to the forefront, so often the listener needs to give some thought to what information that person might really be trying to convey.

Be aware of politics. In telecom environments, as in companies everywhere, there are politics in play. This is very common in changing environments, as people jockey to ensure that their own views and preferences are chosen, and in competitive environments, where people try to position their own value and accomplishments as better than those of others. We have seen that telecom projects have both of these characteristics in spades. Be aware that there could be political motives behind what others say and do, and think a few steps ahead of the others to manage the perceptions and the directions.

Learn about the other people in the project: the stakeholders, the customer, and management. What are their interests? What drives them? What do they expect from the project? Armed with this knowledge, it will be easier to communicate to them information that will keep them seeing the project as meeting their expectations, thus minimizing the pressure they might apply to the project manager and the team. In many cases, it will be necessary to manage their expectations so that they will agree with changes in direction that the team deems necessary for the project.

Be aware of who the stakeholders are, and ensure that the key people always have the information that they need. The time to communicate will usually be less than the time required to explain how your project went off the rails when information was not conveyed or conveyed improperly. Do not assume that people know what is happening with the project, or that they know you are doing a good job. Communicate this.

Be aware also of who you are and to whom you are talking. In telecom, many people working on projects are engineers or IT people. Typically, engineers are introverts who love to work with very complex material, love to know all the details of the material, and, in fact, have difficulty functioning without knowing all the details. Communications often originate with engineers, and also often require the involvement of technical people. When these people are involved, they need to understand the details. On the other hand, many communications related to projects are in place for the benefit of senior managers. Senior managers tend to need a high-level view, and they do not have the time or the inclination to get into the details. They do need to understand the status of the project, the issues, and the action plans. So communications for senior managers, especially when they come from detail-oriented people, need to be carefully designed to ensure that the right information gets to management in a form that they will actually hear and understand.

Be aware of the fact that communication happens not only when you are speaking to someone or writing a note, letter, or report. In fact, a high proportion of communication is nonverbal. Body language, personal demeanor, how one is dressed, accessories, and mannerisms all convey information to the receiver. The image that these things create will affect how the receiver hears and understands the verbal communications. It is wise to review what message is being conveyed by these other aspects of the environment when communicating with others.

Presentations play a large part in the communications for many projects, as team members present to each other, to management, to the sponsor, to the customer, and possibly even to other teams or outsiders such as suppliers. A good presentation does not just happen. The preparation of the presentation must be well planned, and the actual presenting of the material must be done with thought, planning, and, if possible, advance practice. The developer must understand clearly the purpose of the presentation as well as the message. He should set goals to ensure that the right message gets across. The material needs to be both designed and presented in a way that will work for the audience, so it is necessary to understand the audience prior to the preparation, and presentations need to be customized as necessary if the same information is to be presented to different audiences. The approach should be geared to the type of people who will hear the presentation, and the information may need to be augmented or cut down depending on the knowledge that the specific audience already has about the project or issue.

When most people listen, they usually listen for what is in it for them. They might not do so consciously but it does happen nevertheless. The presentation needs to include information that will help the listener understand why he or she is better off with the information being presented, or why he or she should act as the presenter proposes, for his or her own benefit. People preparing presentations generally focus on what they themselves are trying to gain by making the presentation. The audience will usually not care about this at all, which means that the first instincts of the presenter on how to convey the information will very likely be wrong. Know your audience and convey to them what they will gain from your presentation.

Ensure that the presentation has a conclusion, and that this conclusion is presented in such a way that it is clear. Ensure also that the conclusion follows logically from the information presented, perhaps making a point of bringing forward the linkages if needed. If you are selling an idea or material, make sure that the presentation ends with the pitch for making the sale.

In projects, a high percentage of the communication is done via reports. Keep reports to a minimum, and keep the information clear and concise. Include only relevant information in the report, putting any details or backup information into an appendix. If there are references to material outside the report that support the logic or the conclusion, be clear in referencing these. In reports, as in technical papers, make it clear why the work is beneficial to the company, the industry, or the audience. Make the most significant points clearly, highlighting these when possible. If the report describes something that will make work or life better for others, show clearly how the project, the idea, or the work will improve processes, development, or implementation over what exists today.

Finally, one key aspect of communications in projects is that people—team members, management, sponsors, customers, functional managers, suppliers, extended team members, and so on—can all function better, and help the project team better, if they are fully aware of all the relevant information that affects them. Ensure that all relevant information is conveyed, even the bad news. It might be unpleasant to have to be the messenger who brings the bad news, but doing so at least gives those receiving the news the opportunity to determine how they can deal with the situation, rather than having it hit them by surprise.

Hopefully, the information provided here will be helpful to many project teams in structuring and planning future project communications, that is, of course, assuming that the writer communicated it well.

CHAPTER 11

CREATING THE TIMELINES

One of the first questions that sponsors, customers, team members, and project managers ask when a project is undertaken is when the final deliverable can be ready. Creating the schedule to determine the project timelines is one of the most critical tasks for the project team, and one that everyone wants to do as early as possible. No project implementation should begin until the schedule has been determined. Armed with a schedule, the work can move forward, flowing smoothly with each activity occurring at an optimal time for the project. As we have seen in the preceding chapters, it is necessary to first clearly define the scope, and break it down into activities so that a schedule can be built that includes all the project activities. Once the WBS has been created, the building of the schedule can start.

There are two tasks that need to be completed before the project schedule can be determined. Either of these can occur first, since neither is dependent on the other, but the schedule requires the completion of both. These are determining the duration required to complete each activity at the bottom level of the WBS and also determining the dependencies between and among the project activities.

TASK DURATION

We will address the durations first. This is generally done by asking the person who will complete the task how long it will take to do that work. In order to accomplish this, it is necessary to assign the task first, and also to have access to the responsible person. Although tasks are not always assigned to specific individuals at the beginning of a project and, indeed, sometimes the specific individual who would be available for a task may not even be known, the plans can be more accurate if this information is known, as the duration of a task will vary considerably depending on the skills and experience of the person doing it. Therefore, it is best to first determine, as much as possible, who will be assigned to each work package (bottom level element of the WBS). Then the duration can be estimated for each task. If the assigned person is known, the estimate will reflect the time that it would take for that individual to do the task. If the assignment is not known, the planner can use an average, common, or standard time for such a task, especially if it is a task that has been a part of previous project, and so well understood. Regardless of who does the estimates, it is also wise for the project manager to know or learn about the accuracy of past estimates made by this person, as some people consistently overestimate or underestimate the time it takes to do something. If the estimator has one of these tendencies, the PM can work with this person and make modifications after the fact to ensure that the estimate is as accurate as possible. Similarly, when historical data, industry standards, or averages are used, it would be best to think about whether or not there are differences from the average for the specific task, and possibly modify the estimate to compensate for these.

DEPENDENCIES

Next, we need to determine any dependencies any one activity has on any other. If no activity relied on the all or part of any other activity, it would be possible, with enough resources, to start all activities on the first day and work through all of them in parallel. Of course, this is unlikely because for most projects not enough resources have been assigned to allow this possibility. But if it were to happen, the length of the project would be determined by the length of the longest activity. Since in any realistic case, at least some, and probably most, of the activities will be dependent in some way on others having been finished or at least started, it will be necessary that some activities occur earlier than others, creating a sequence of activities within the project. If one activity needs another to have been started or finished before it can start or finish, we say that the first activity is dependent on the second, or that there is a dependency between the two activities. We need to determine all of the dependencies of any activity and any other one. Once all the dependencies are known, it will be possible to line up the activities in order, thus creating a structure for the optimal way in which to tackle the work. In some cases, one activity might be dependent on more than one other activity. However, it is best initially to work with the activities in pairs, identifying where the dependencies are and in what way each is dependent on the other. If all dependency pairs have been identified, then those multiple dependencies will appear, as they will consist of multiple pairs with a common end point.

But in order to identify these dependencies, it is necessary first to understand the types of dependencies that exist. There are four ways in which one activity can depend on another, as illustrated here. Let's call the two activities A and B in each case.

Finish–Start Dependency

Activity B can start when Activity A finishes. But B depends on the finish of A to enable it to start. So B cannot start until A finishes. This is a finish–start dependency or FS dependency. This is illustrated in Figure 11-1.

When A finishes, B can start. It is the completion of A that allows for the start of B. Thus, when we have reached the state in which A has been completed, B can then start. But B cannot start until A finishes.

Start–Start Dependency

When A starts, B can start (Figure 11-2). It is the start of A that allows for the start of B. Thus, when we have reached the state in which A has been started, B can then start. But B cannot start until A starts.

Finish–Finish Dependency

When A finishes, B can finish (Figure 11-3). It is the completion of A that allows for the completion of B. Thus, when we have reached the state in which A has been completed, B can then finish. But B cannot finish until A finishes.

Start–Finish Dependency

When A starts, B can finish (Figure 11-4). It is the start of A that allows for the completion of B. Thus, when we have reached the state in which A has been started, B can then finish. But B cannot finish until A starts.

Note that in each case we say that the successor activity *can* start or finish, not that it *must* start or finish once the enabling state has been reached. The requirement is for the enabling state to be reached, not for the successor to take any specific action.

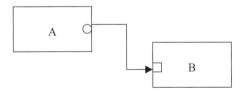

Figure 11-1. Finish–start dependency.

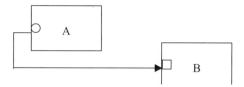

Figure 11-2. Start–start dependency.

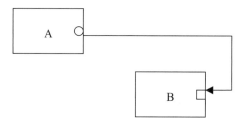

Figure 11-3. Finish–finish dependency.

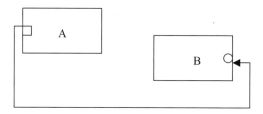

Figure 11-4. Start–finish dependency.

Note also that we are describing something that *needs* to occur, not just stating that something *did* occur. The fact that something has occurred, or did occur, is nice information but it does not imply that there is any dependency between the action that has occurred and something else. When we state dependencies, we need to be clear in stating what the requirement is, and what that requirement being met will enable.

Note also that there is no dependency that says that two things must happen at the same time. In a start–start dependency (SS), it is possible that A and B will both happen at the same time, but that is not necessary because of the use of "can" rather than "must." If there is a requirement for two things to happen at the same time, then there must be two separate dependencies that occur simultaneously. For example, we could have a situation in which A cannot start unless B has started and also B cannot start unless A starts. This double dependency would then cause A and B to have to start at the same time, and thus they would happen to occur in parallel. But it takes two separate dependencies to cause this to be the case.

So let us think of examples of each of these in order to illustrate how each works.

Finish–Start

Finish–start dependency is by far the most common of the four. When people think of dependencies, it is usually finish–start dependencies that come to mind. Some possible examples are:

Only when the network design is finished can we start installing the circuits.

When the evaluation of the responses collected on the customer feedback sheets is complete, we could start to design the new service.

When the installation of the call center computers has been completed, the training of the operators can start.

Start–Start

Some examples of this dependency include:

When the phone system in the call center starts to work properly, the agents can begin taking the calls.

When the manufacturing assembly line starts to roll, we can start to assemble the handsets.

Finish–Finish

Examples of finish–finish dependencies are:

We cannot finish installing the drywall in the test center until installation of the wiring finishes.

We can finish the contract negotiations when we finish identifying all the local arrangements requirements.

Start–Finish

Start–finish dependencies are always the hardest to think of because they are the least prevalent of the dependency types, but there are many of them nevertheless. One excellent example is:

When the night shift worker starts to work, the evening shift worker can finish. However, if a snowstorm or some major disaster prevents the night worker from arriving at the office on time, the evening shift person will have to stay and work to prevent the station from being unmanned. But once the night shift worker starts, the evening shift can finish.

MANDATORY AND DISCRETIONARY DEPENDENCIES

The descriptions of dependencies do not end here, however. There are additional characteristics that might be part of any of the four dependency types. First, any dependency might be either mandatory, or a hard dependency, or discretionary, or a soft dependency. A mandatory dependency is one that cannot be ignored or broken. For example, we cannot begin to send data over a private line until the completion of the installation of all the components of the line. No matter how much we might want to send data, this is not possible until the installation of the line has been completed. This is a hard dependency. A soft dependency is one that is in place because it is good practice to do this, or a rule that we should follow, or the way in which the company prefers to do something. If needed, discretionary dependencies can be ignored. Since these exist for good reason, they should always be built into the project design, and they should always be honored unless there is some significant issue to cause them to be dropped. If the project just cannot be completed in time with all the dependencies in place, it is possible to drop some of the soft dependencies. However, it would be preferable to keep these in place during the planning period if at all possible. As the work progresses, many projects start to run behind, and people start to look for ways to make up time. If soft dependencies need to be dropped, this would be a better time to waive one of these.

LAGS AND LEADS

The other characteristics that might be inherent in any dependency are lags and leads. Lags are delays in the start or the finish of the successor activity following the relevant start or finish of the predecessor activity. In each case, as above, these are part of the dependency itself, not just something that happens. So if there is a dependency between two activities that includes a component that is an advance or a delay, the dependency includes a lead (advance) or a lag (delay). Leads and lags can occur with any type of dependency. Some examples could be:

Finish–Start. Three days after the construction team that is building the new data center finishes pouring the concrete in the foundation the construction of the building frame can start. The dependency is finish–start, but it is not a direct finish–start because it is necessary to wait for 3 days for the cement to cure before starting to build the frame. If this delay is not respected, and frame construction starts earlier, the risk that the building will later collapse is very high. So the delay, or the three-day lag, is part of the dependency itself.

Start–Start. Two hours after we start to download the data regarding network performance, we can start to use the data to build the diagram showing the network performance and we can start to analyze the performance. In fact, perhaps the preferred mode of operation is that we complete the download of all the data before we start to draw the diagram and do the analysis. But maybe there is not really enough time to wait for the full 10 hours of the download and still meet the deadline. And suppose that it has been determined that in the first two hours the majority of the critical data for the backbone portions of the network are given, with the remaining hours producing specific data on the different local legs of the network. So the finish–start dependency can be there, but it is a discretionary dependency, there because we prefer to wait until we have all the data before we start working with it. But there is also a tighter dependency that is a mandatory or hard dependency. Once we have the backbone portion we can start the diagram of that portion of the network and start analyzing that portion of the performance. So there is a hard start–start dependency with a two-hour lag that we need to respect. We do not need to start working with the data after the two hours, but we cannot start until we have at least that portion of the data that downloads in the first two hours.

There are many such examples, and these can occur for any dependency type, as long as the lag or the lead is inherently part of the dependency itself.

PROJECT LOGIC DIAGRAM

Once all the dependencies have been determined, the network diagram can be created. In telecom, people draw diagrams of telco and customer networks, showing equipment and facilities connecting the equipment and offices, and call these network diagrams. In project management, we also have network diagrams, but these are quite different from the telecom ones. They show the project activities, and the flow in which they should occur in order to complete the full project. These networks can also be called a number of other names such as project network, logic network, logic diagram, and so on. By any name, we have the same network.

To create the schedule and show the flow of the work, we sequence the activities. At some point, either before or after sequencing, we also must identify the duration for every activity. These two components—duration estimation and sequencing—plus defining the actual work items comprise the building of the schedule. The duration is the length of time it is going to take to do the activity. Some things are very difficult to predict. Some new product development activities, never having been done before, are inherently impossible to schedule exactly, but the best example of schedule uncertainty I can think of comes from agriculture, not engineering, for if there is anyone who has to deal with uncertainty every day, it is a farmer. Imagine the stress in scheduling a harvest. Extremely expensive equipment needs to be leased well in advance, and the farmer has only statistical information on when the harvest can begin. Once begun, the duration of the harvest is at the mercy of the autumn weather, because you cannot harvest wet crops. In a situation like this, there are inherently unpredictable elements, outside the control of the task owner. There are lots of activities that are hard to predict. People who work in those areas that are hard to predict but are not completely novel do have ways to estimate the time for these, because they handle similar activities every day. Using past data on these thousands of experiences, they have worked out an approximation for the duration of a typical occurrence of the activity. For example, how long does it take to solve a problem when a trouble call comes in? Statistically, you can get some numbers but we still do not really know how long any individual call will take when it arrives. You can only give the statistical average or typical time. For our projects, we need to put some kind of estimation on the duration for every activity that is included, even though some can be extremely difficult to predict. Generally, people do use statistical numbers for the items that are highly unpredictable.

Sequencing is essentially just the application of all of the dependencies that have been identified. One way to start could be to assume that every activity will start on day one of the project, with everything happening in parallel. Then the dependencies can be applied, one by one, causing activities to move to later start dates as they wait for other activities that must start or finish before they start or finish. If all the actual dependencies have been identified in the dependency work, then an optimal network will result. The network will show a number of paths— sequences of activities—that must all be completed in order to complete the full project, because all activities must be included in order to finish the project. If the resulting network has some paths that just end, leading to nothing, the team must start asking questions. Is this activity the last action that completes the project? If so, that is fine, but if not, why is it that nothing else in the project is dependent on the last activity in the path? Do we not actually need the items in that path in order to complete the project? If they are not needed, why are they there? If they are needed, what is their benefit, and how do they benefit the final outcome of the project? If it does affect the outcome, there must be a dependency of some other item on this path, so the end of the path should feed into the activity that depends on it. These hanging tasks can be dealt with one by one until there is a fully connected network for the project.

Let us look at some examples.

Figure 11-5 shows one sample of a schedule that starts with a single initial task and completes with a single completion task. The work in the project is done sometimes in parallel and sometimes not. The project is not very complex, as the number of tasks is quite small, but it is excellent for illustrative purposes to show a representation of the schedule.

In this case, the representation uses the precedence diagram method, or PDM, in that the activities are shown as nodes which are interconnected by arrows to show the dependencies. As long as a dependency is finish–start, only the arrow is needed. However, if the dependency is one of the other types, then letters identifying the dependency type must accompany the arrow. And if there is a lag or a lead associated with any dependency, then a positive or negative number to indicate the size of the

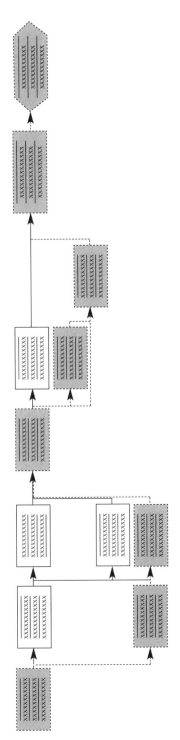

Figure 11-5. Engine schedule.

lag or the lead must accompany the arrow. So, for example, if a dependency were finish–finish with a lag of 10 days, the arrow must be labeled with FF +10.

Of course, most diagrams will be more complicated than this one. One sample, partially shown in Figure 11-6, has quite a few more items than the engine schedule, but we can still see the flow of all of the activities and understand how the project work will occur.

CRITICAL PATH

Looking at these sample diagrams, we can see that most of the boxes, or activities, are shown in black, whereas a few are shown in gray. There is a reason for this. Each of the activity sequences is a path through the network, and as mentioned above, all activities must be completed, so all paths must be completed in order to complete the project. If we consider all the paths for a specific project, then we could look at the durations of all the activities within each path, and we could calculate the length of time that it will take to complete each of the paths. Since all paths must be completed in order to complete the project, and since every path starts with the initial project activity and ends with the final activity, then the path with the longest duration will define the shortest time within which the project can be completed. There is a name given to the longest path because it is so important in determining the length of the project; the critical path. The reason for this name is that it becomes critical that none of the activities on this path be delayed or the project schedule will be in trouble. Once the project's sponsor and stakeholders see the time lines for the project, they see the date by which this longest path finishes, and they then expect that the project will complete by this date. Dependencies and risks may get forgotten, but that final date will always be remembered! Paths with a shorter total duration will be able to tolerate some degree of delay without pushing out the project's completion date.

Let us look at the possible paths through a network and calculate the length of each path. In the network in Figure 11-7 there are 4 paths:

ABF
ACF
ADEF
ADEGF

Since some of the dependencies are not finish–start, we need to calculate the path lengths carefully. The durations of the tasks are shown in the boxes in the center bottom. It will take $3 + 10 + 9 + 5 = 27$ days to complete ABF, because once we finish B we must wait nine days before we can start F.

It will take $2 + 16 + 5 = 23$ days to complete ACF. We use two for the duration of activity A because the start–start dependency allows us to start C once we have completed 2 days of work on A.

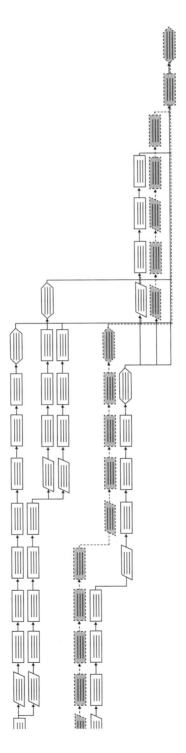

Figure 11-6. More complex network.

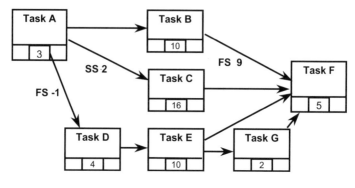

Figure 11-7.

ADEF will take 3 – 1 + 4 + 10 + 5 = 21 days, and ADEGF will take 3 – 1 + 4 + 10 + 2 + 5 = 23 days.

Thus, the critical path is ABF, since we are likely to promise the end result in 27 days, and if that happens, we cannot afford to be late with any of the activities on the ABF path.

In addition to knowing which activities we must monitor most carefully, we also need to know how early we can start any specific activity in the project, and how early each activity can be completed. We can determine all the dates for starting and finishing all of the project activities by two techniques known as the forward pass and the backward pass. Completing these will also allow us to understand how much flexibility, or float, there is in the handling of each activity in the network. To illustrate these techniques, consider the simple network diagram shown in Figure 11-8.

Forward Pass

We will look at the forward pass first. This is the technique for determining how early each of the activities can be started. It is best to work on this without considering the calendar, so that the optimum flow can be determined first. Once we have the flow, we can then add the calendar to ensure that all of the activities can actually be started on the dates on which they fall, and that the work can be done smoothly through the dates. There will often be reasons that will preclude this, so it is best to determine the flow first, and then deal with these later. It might be necessary, for various reasons, to change the network, and if that happens, then any previous calendar considerations would have to be redone. To avoid this rework, and also to avoid the distraction from the task at hand, it is best to determine the optimum early start and finish dates first, and then later work the network into the calendar. So we will work first with general information on the timing, starting and finishing activities on Day x or Day y without involving any specific dates.

It is also necessary to make some assumptions about the work day in order to be consistent in the timing of the activities. We can assume that if we start a task on Day x, that we would start working on that task first thing in the morning of Day x, but that if we are to finish a task on Day y, we can actually finish it by the end of the workday on Day y. These are fairly common assumptions, so most people find that they work well. If these assumptions do not work in some environments, the people working there could make other assumptions, such as the availability of additional shifts. But we will use these default assumptions for discussing the forward and backward pass.

Consider the diagram in Figure 11-8. Let us assume that we start work at 8:15 in the morning, so our early starts will start at 8:15 on the specified day. Let us also assume that we finish work at 5 pm each day, so the early finish times will be 5 pm on the specified days.

This diagram is not the same as the one in Figure 11-7, but it does have four paths through the network. We will walk through each of the paths, from the beginning of the project until the completion, and calculate the earliest day on which each activity can start and, thus, the earliest day on which each can finish. This is the forward pass.

First, the paths are:

ABG, which is 19 days long
ACG, which is 23 days long
ADEG, which is 27 days long
ADEFG, which is 27 days long

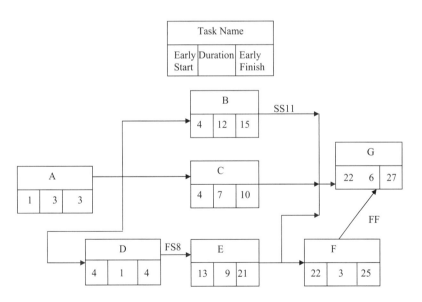

Figure 11-8.

A is the first activity, so we can start it on the first day. It has a three-day duration, so we would then finish it at the end of the third day. These numbers are shown in the squares along the bottom of the activity diagram.

Once we have completed A, we can start B, C, and D, so each of these has a 4 in the box showing the earliest start day. Since we will finish A by the end of Day 3, we will be able to start these three activities the morning of Day 4.

Activity B would then finish at the end of Day 15, C at the end of Day 10, and D at the end of Day 4. And we can start E eight days after we finish D, which is the end of Day 12, but since we do not start working at the end of the day, work would start on the morning of Day 13. If we had only one path, ABG, then we would be able to start G seven days after the start of B. However, we will not be able to start G this early, because the other paths will require that it start later. If we had only ACG, we could start G on Day 11. But before we decide to do this, we need to check the remaining two paths. In ADEG, we find that we can start G only on Day 22. For ADEFG, we find that we need to complete F before we can complete G. F completes on Day 25, so as long as we finish G on or after the end of the day on Day 25 we are fine. If we start it on Day 22, this will be the case. So the earliest that we can start G is Day 22.

Completion of the forward pass shows us that this project can be completed, at the earliest, by the end of Day 27. In the real world, though, things will not be this simple. First, we will need to add the calendar to ensure that we do not lose days within this 27 day period due to holidays, vacations, important meetings not related to the project, and so on. Also, once we have the full network information we will need to ensure that the people or other project resources are not double-booked by this schedule. And we also need to ensure that we include contingency in our network, which we will do as well once we know where to put it. So for the moment, Day 27 is only a working number, which will likely change before the planning is done.

Backward Pass

The next step is the determination of the latest dates we can complete each of the activities without compromising the project end date. For the moment, we will use Day 27 as the project end date, as this will allow us to work backward through the network and determine the latest completion date and, hence, the latest start date for every activity in the network. Walking backward is similar to walking forward through the network, except that the focus is mainly on the end dates for the activities rather than the start dates. Start with G, and since we already know that it will end on Day 27, we can see immediately that the latest start date for G is Day 22.

Then since F can finish only when G finishes, the latest finish for F would be the end of Day 27, which gives us a latest start of Day 25 for F. Working backward along this path, ADEFG, we would then find that we can finish E on Day 22. We could have finished E later if we had only F to worry about, but since we must also complete E before we can start G we are constrained to finishing by the end of Day 22. That makes the late start for E Day 13, and ties the late start and finish for D to the same dates we found in the forward pass. A also remains unchanged from the forward pass, because of the constraint placed on it by D.

Walking back the middle path, we see that C does not have to complete until the end of Day 22, so the late start day for C is Day 16.

And on the top path, the BG dependency is a start–start, so the constraint here is on the start of B. It can start as late as the beginning of Day 16, and then it would finish at the latest at the end of Day 27, which is still fine because this is within the working time of the project.

Float

It is clear, once this point has been reached, that the team has some flexibility in the timing of some of the tasks, but that others must occur at fixed times if the project is going to be able to complete on time. The flexibility that is available to some activities is called float. In our network, the critical path is ADEG. The activities B, C, and F all have some float, and these activities are not on the critical path. In fact, if the project completion date is tied to the early finish date for the last activity, as was the case in this example, none of the activities on the critical path will have any float. If the due date is later than the earliest finish date, then the critical path will have float; the float for the path in total will be the same as the number of days that exist between the early finish date for the project and the due date. So if we could finish a particular project by Day 110, and the due date is Day 123, then the critical path will have 13 days of float, because it is acceptable to slip the final activity by 13 days and still finish by the due date. The float on the other activities will also be greater than it would have been without the 13 day window at the end.

The float in out network shows in the three activities B, C, and F. According to our calculations, B has 12 days of float because it could start as early as Day 4 or as late as Day 16. Anything in that range is fine. C also ended up with 12 days of float, because it can finish as early as Day 10 or as late as Day 22. F can finish anywhere between Day 25 and 27 so it has three days of float. Therefore, if we have problems with one of these activities, or if we need to move one of them because of resource problems, we can make changes as long as we stay within the identified windows.

Please note that float is not something that can be controlled or modified by the project manager or the team. Once the project due date has been set, there will either be float for some elements or not. This is a function of the positioning in the network, and whether the path for a given element is the critical path or a shorter path. In a noncritical path, some elements will have float, because the path is shorter than the longest path, so there is extra time for the work on that path.

Showing the Schedule

Now we are ready to show the actual schedule of the project. Once we know the actual start date or end date, we can fit the network into the calendar, taking care not to break any of the constraints placed on the start and finish dates. If this overlay on the calendar works smoothly, we will have the schedule. If there are problems, such as having to shift everything to one day later due to a calendar holiday, or maybe five days later due to company mandated vacation days, we need to look at the impact of these. Moving activities later will use up float, and if there is no float, this will necessitate moving the finish date later. Trying to move things earlier in time is much more problematic, and we will address this shortly when we discuss collapsing the schedule.

The schedule can be shown in many different formats, and for some companies different formats are used for different purposes. The two most common formats are those that are produced in MS Project: the project network and the Gantt chart. The network looks like the ones shown in Figures 11-5 and 11-6, except that it is usually laid out on a page that has a calendar running along the top or bottom, and the tasks are shifted to have their start dates line up with the actual calendar start date. Since all boxes in the network are usually the same size, the finish dates are not illustrated diagrammatically, but they are usually written on the activity so that all of the activity information is available in one place. A sample is shown below in Figure 11-9.

The second common method of showing the schedule is the Gantt chart. In this format, the activities are also shown with arrows illustrating the dependencies, and with a calendar along the top or bottom of the diagram. However, in this format the task boxes are elongated to form bars that illustrate the length of the tasks. So a longer task will appear as a longer bar, and both the start and finish dates of the tasks can be read. The Gantt chart in Figure 11-10 shows a portion of the same project shown in Figure 11-9, except that the Gantt starts about one week earlier in the project than in the network diagram.

Both of these formats are clear, illustrative, and useful. The network diagram is a very good planning tool, as it makes the task dependencies and critical path very clear. The Gantt chart is generally considered a better tool for tracking the status of an ongoing project.

Including Contingency

As mentioned above, we must include extra time in the schedule, called contingency, to allow for recovery from problems that might occur. The amount of time is something that should be calculated, and the methodology for determining the amount of contingency time to include is shown in Chapter 9.

Here, we will assume that we know the amount of time that we need to include for contingency, and we will address how to incorporate that time into the schedule.

The contingency is extra time that is incorporated into the schedule to allow the team to be able to recover after something has gone wrong. If we need to include

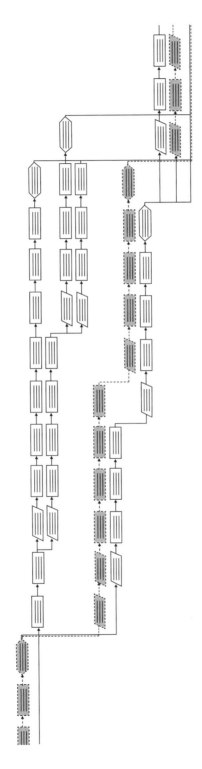

Figure 11-9. Network diagram for project A.

135

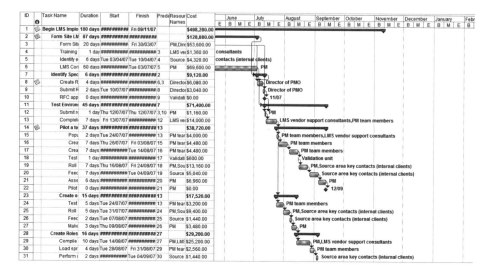

Figure 11-10. Gantt chart.

20 additional days for recovery, these could be placed at the end of the schedule by quoting a completion date that is 20 days later than the date calculated via the network development. In this way, all the contingency would be in one place, at the end of the schedule, and whenever something slipped in time, the final completion date would move closer to the promised completion dates. Although this does allow for minimal disruption to the schedule plan, and lower risk that any specific activity will use contingency when it is not really needed, it is not a good way to handle the contingency. First, it is far too obvious to everyone that there is a lot of open time at the end of the project, so management or the customer is very likely to decide that the project completion date should be moved up. Second, the team might resent being pushed to complete tasks on time according to the original schedule when they see the big pot of time sitting open at the end of the project. And finally, since it is very likely that this time will be needed in bits and pieces throughout the project, if we do not incorporate it earlier in the schedule, then as soon as one item on the critical path slips, the timing will be off for all subsequent items. So the project will always appear to be off track, and if there are resources that must be scheduled for specific dates, these schedules might need to be renegotiated multiple times as the schedule keeps running further and further behind the initial plan.

So it is better to incorporate the time into the schedule itself, in chunks. The question is, where is it best placed? The first consideration is the critical path. Placing contingency into items that are not on the critical path is less useful than placing it on the critical path, where it is most likely to be needed. Other tasks off the critical path already have some degree of float. Tasks on the critical path typically have no float.

Contingency time is required because there are risks to the project work that might materialize. If a risk does materialize, then more time will be needed for either the activity that encounters the risk or those following this activity, so the contingency would be well placed in these risky activities, as it is most likely to be needed there. Also, if there are many paths feeding into one particular area of the critical path, then there are many chances that that point could be impacted, so it would be wise to add some of the contingency at that point.

The project manager and/or the scheduler needs to consider the actual project and its activities to make intelligent decisions on the best places to put contingency for the project, and how much contingency to place in each position.

Collapsing the Schedule

Once everything has been incorporated into the schedule, the team can step back and assess where things stand. With luck, the due date, which is usually set externally, will be at or later than the actual finish date shown in the schedule that includes the contingency. If so, then the last step, a detailed check of the loading of each of the resources, can be done to give final assurance that the plan is workable.

Unfortunately, this very rarely happens. The people making the determinations of the duration of the individual tasks in the schedule are motivated to be conservative in the durations they report, as they will be held to meeting these durations, so a complete schedule often does not meet the desired end date. This is one main reason why the PM is encouraged to discuss the actual time requirements with each team member, to remove any contingency that they have included from the actual work time estimates, but, of course, to build the project contingency into the schedule in an optimum way. Then, if the schedule completion date is later than the due date, this problem must be addressed before the schedule can be completed. If the schedule is too long, it is necessary to find ways to shorten it.

There are a few things that can be done. These include shortening the critical path by adding resources to shorten some of the longer tasks, ignoring some of the dependencies to allow tasks to be completed earlier than the time they would naturally be completed in the schedule, and shortening the critical path by dropping some of the activities on the critical path, or shortening these activities by reducing the quality of the end products.

The first technique, shortening the critical path via the use of additional resources, is called *crashing*. The critical path can be shortened if some of the activities on the critical path can be shortened. When the activities are effort-driven activities, which we will explain shortly, this can be done by having the people who are assigned to the activities work longer work days, which generally also requires that they be paid overtime, probably at a higher hourly rate than their usual rates. In some cases, including the overtime will be enough to shorten the critical path sufficiently.

However, if that is not enough to produce the required results, it might be necessary to add additional people instead or as well. Of course, for most activities, assigning *n* people to do the work will not reduce the required time to 1/*n* of the total required time, especially for the type of activities required for telecom projects. These are usually highly technical and/or professional, so skill is required and trying to have more than one person working can create the need for additional communication and co-operation. Adding people can help in many cases, but the reduction in time will not likely be linear. The focus of schedule crashing must be on the critical path, since shortening other tasks will not help to complete the project earlier.

The next technique for reducing the time of the project is called *fast tracking*. With this technique, some of the dependencies are ignored to allow some of the activities to be completed earlier. It is not possible to ignore mandatory dependencies, but discretionary dependencies are in place only because of preference. Generally, when one ignores a discretionary dependency some time can be saved, but the dependency cannot be removed completely, because at some point a hard dependency will still apply. For example, good practice might suggest that programming should be completed before testing begins to avoid having to retest segments that might be changed as the development proceeds. The discretionary dependency is that programming must finish before testing starts. This dependency might be ignored to save time, and testing of modules, sections, or specific portions of the code can be tested before all the programming has finished, in hopes that these tested segments will not have to be changed. However, it is not possible to test until at least some portion of the code has been written. So there is a hard dependency that will control the amount of time that can be saved by the fast tracking.

Another technique for shortening the critical path is dropping one or more activity on the critical path. This can be helpful, but it is important to keep in mind that the activities are there for a reason, and dropping them will either cause some portions of the scope to not be delivered at all, or to be delivered in a state that does not meet the expected quality. Therefore, if this is to be done, the stakeholders who are looking for the affected deliverables need to be informed, and their expectations must be readjusted.

The last technique involves doing everything, but not doing it as well as was initially planned. This will also have the same impact on the deliverables and, therefore, it again requires discussion with the affected stakeholders.

Applying any of these techniques will change the balance of the network, which will require a reassessment of the resource allocation. But before discussing this, we need to ensure that the nature of the tasks is clear.

EFFORT-DRIVEN AND DURATION-DRIVEN TASKS

Earlier, it was mentioned that tasks can be shortened by adding resources only if the tasks are effort driven. Effort-driven tasks are tasks that require a certain amount of effort in order to complete. Examples could be testing facilities, preparing customer presentations, or writing questions for a test.

Some tasks, though, are duration driven. A duration-driven task takes a certain period of time, no matter how many people are assigned to it. Some examples of duration-driven tasks are a 24-hour test or three-hour baking of a component. If the activity must run for a certain length of time, it must run for this time regardless of the number of people working on it.

AND THEN . . .

Once the final schedule has been manipulated to a form and length that is acceptable, it is important to check the resource requirements over time to ensure that the resources available match the ones needed. This is done by looking at the network through every time period (weeks, days, or months) and determining what resources are needed during that period. The analysis can be done for each individual resource through the project and also for the project as a whole. If all of the project resources are available for project work full time for the duration of the project, it is important to distribute the work in such a way that each person has work throughout the project, and to ensure that the number of resources required matches the ones available at all times. If this cannot be achieved, then there will be idle resources (who are being paid from the project budget to work on project activities) and/or overloaded resources. The better the match, the more efficient the project resource usage will be.

If there are problems with the loadings, it might be necessary to move tasks around again or to reassign people to achieve the desired results. Of course, the moving of anything requires a reassessment of all the considerations discussed above. This will be an iterative process, hopefully converging to an effective solution.

Most people will use some project management software application at this point, to assist with these modifications. All the network design, resource leveling, and network restructuring can be done manually. However, this is a lot of work, and programs such as Microsoft Project, which seems to be the most commonly used application, can help considerably with the work. It is not the intent of this book to recommend any one application, or even to discuss applications. However, it is recommended that software be used to assist wherever possible, which can then allow the PM and the team to use their time and energy in other aspects of the management. These applications cannot replace the people in the teams. For example, MS Project can design the network, but only after the team has determined the work packages and their durations, and especially the dependencies amongst them. The program can work with these, but it cannot determine them. Similarly, the people must determine the work assignments. Given these and the information above, the program can determine the resource loadings, showing the PM where there are problem areas that the people must resolve.

Overall, the creation of the time lines for a project is an iterative process. The team must determine the activities that must be done, assign each to a team resource, and determine the cost of each resource and the duration needed for each activity. The project network can then be created, and a calendar can be overlaid on the network once the start and/or finish dates are known. The team will need to change the network to resolve any problems, such as date constraints that are not met, resource loading that does not match the resources available, and even changes that occur as the planning and work move forward. Each time the network is redesigned, all the above items must be checked again to ensure that no new problems have been created. In the end, the resulting schedule can be used to help manage each and every activity so that everything can be completed on time.

CHAPTER 12

MANAGING THE COSTS

For most projects, the stakeholders focus most strongly on the time to complete the project and the overall cost. Telecom projects are no different. In fact, given the rapidly changing environment, with the competition coming from every direction, these are more intensely monitored than they are in many other industries. Costs were always an issue for telecom projects because the networks are large, each technology is expensive since it is extremely complex, and a large number of complex technologies are needed to create a viable telecom service or product. In the past few years, as the rate of introduction of new technologies has increased rapidly, the costs for selecting technologies and designing, developing, and implementing new networks, services, and products has also escalated due to the need for people to be constantly evaluating new product and technology offerings to keep their technical skills up to date and to assess and replace technology on shorter timescales than in the past. The need to discard technologies before they have been fully depreciated has raised the costs of doing business in this environment, with a resulting closer focus on the cost of all projects. It is critical, therefore, that people in telecom have a good understanding of the way in which costs impact projects, and the implications to the project team. It is more important than ever that these costs be properly planned, monitored, and controlled.

In this chapter, we will first give a short overview of types of costs that can be incurred in projects. Clearly, there are many financial concepts that could come into play in projects, and these all need to be properly managed. Most of these are required for management in general, as well as in projects. To cover all such concepts properly would require far more space than is allocated to the project cost information in this book. The goal in this chapter is to cover a few cost concepts that come into play in most projects, not to explain finance in general. Other than a brief mention, we will not discuss things such as balance sheets or profit-and-loss statements, which are so prevalent in every business but which are not always as useful for projects. We will describe certain concepts and their relevance in a project environment, but not always give a formal definition.

The main cost item on the minds of project managers, of course, is the project budget. The PM must first determine the amount of money that is available to undertake the project, and then, after looking into the deliverables and work to be done, ensure that the project can actually be completed for the available funding. Once the decision has been made to go ahead with the work, the PM then must ensure that the team adheres to the budget for each item in the project. In addition to this, many PMs must also make reports and/or provide input for corporate accounting, which might go beyond the information needed directly for the project but nevertheless are very important for the company.

We will look at a few of the different types of costs, at estimating the costs for a project, at some issues with making estimates, at project budgets, at profitability measures such as payback period and NPV, and at some concepts that apply within projects, including planned value and actual cost.

In some projects, an income statement or profit-and-loss (P&L) statement might be useful. This statement looks at the financials over a period of time, to determine whether or not the company made or lost money during that time. The total revenue or incoming money received during that period is decremented by the expenses to show either a positive or negative bottom line. This may be relevant for some projects and not a useful tool at all for others. For a project that has been set up to develop a new product or service, the expenses of the development will all occur during the lifetime of the project, but the revenue received from the sales of the product or service may not appear until late in the project or even after project completion. In this case, the income statement is a useful tool for the business overall but not really useful for the project itself.

A balance sheet, which summarizes the company's financial balances at a point in time, may also not be particularly useful on many projects. On the other hand, a cash flow statement might be very useful for the project, because cash flow has the potential to impact what can be done in a project. This statement shows the money coming in and the money going out. For a project, this could be used to compare the money that is provided to the project by the sponsor or other outside sources, and the money spent. The timing of the cash flow might be even more important in some projects than the bottom line for the cash flow over the project life.

Consider a project to organize and run a conference on broadband wireless communications networks. The sponsor of the conference will probably provide some seed money to the conference organizers to allow them to cover their expenses. When a conference is organized by volunteers, as is done for IEEE conferences, the people provide their time and effort free, but anything that must be purchased has to be paid for somehow. The seed money provided by an organization such as IEEE will not cover all the up-front conference expenses such as printing and distributing fliers, having organization committee meetings, sending someone to board meetings to report on progress, paying deposits for hotel or conference space, and so on. The conference is probably expected to make money by the end of the conference project lifecycle, but the revenue that it generates does not start to appear until quite late in the project, after the fliers have gone out to announce the conference and commitments have been made for space, and other items. Therefore, the organizing team must find other sources of funding who will provide cash early in the project life cycle. This is usually called something such as sponsorship, and amounts collected will be part of the overall financial statements. In the statements, the timing of the receipt of this money is not important, but in the project life cycle it is critical, and late receipt of funding has the potential to break the project. The same could be said for new companies receiving funding such as grants for a project. In a project, the actual timing could be far more important than the overall bottom line. The bottom line for the project during the project life cycle could also well be negative, even though the project will produce something that might make quite a lot of money over the product or service lifetime.

TYPES OF COSTS

Many types of costs can be incurred in projects but, generally, the charges that will accrue to a project will include the direct costs of the people or other resources that are used in the project work, as well as any indirect costs, also called overheads, associated with these resources. Direct costs include the salary or direct purchase price of any material. Material costs can be capital, paid to purchase assets that are not used up in the course of the projects, or expense, paid for expendable items. Indirect costs include items such as furniture, phones, computers, benefits, vacations, and the salary costs of people associated with the operation of the company and necessary to the project, although not charged directly to the project. An important concept is the idea of the "loaded labor rate." This is the total cost of the labor of a person on the project team expressed on an hourly or weekly basis, and is the amount that is actually charged to the project budget. This rate comprises the person's actual salary, plus all the overheads as described above. Depending on the company, the loaded labor rate might be two to five times the person's actual salary. There are probably few project managers who have not heard "If only I actually made that much" in discussing the loaded labor rate.

Sunk costs are amounts of money that have already been spent. Once any work has occurred on a project, there are some sunk costs. These costs increase as the work progresses, and will hopefully equal the planned project budget at the end of the project. Sunk cost is often confused with equity, and teams often feel that because the company has already spent a large amount of money on their project, it will go to completion regardless of cost overruns. This kind of thinking often results in throwing good money after bad, and it is necessary for management to continually monitor the viability of a project, and if that viability is no longer there, to terminate it. As an example, suppose that a company has been building a separate system for taking orders for some new product line, because the existing system could not accommodate the information needed for this product line. The system is being built in-house so that it can be designed to include all the required name, address, equipment, rate plan, and expected usage information that is needed. When the new system has been almost half completed, senior executives decide to replace the current order system with a completely new one, and this new one can be designed to accommodate the data required for the new product line. The new overall system will be available three months later than the product line system can be completed. However, some of the resources who are working the product line system will be needed to work on the new overall system and this will cause the product line system to be even later, probably by 2–3 months. At this point, it no longer makes sense to continue building the product line system, even though more than half of the original budget has already been spent, and there is very little that could be used for other purposes at this point. The money already poured into the product line ordering system is sunk cost.

Opportunity cost is a different way of looking at cost. Rather than being a measure of the actual money spent for a project, it is a comparative measure of cost. One definition specifies that it is the cost of benefits foregone in order to do the project. Thus, if we decide to offer VOIP service over a packet-switched network, which prevents us from growing our wireless service portfolio, then one cost of building the VOIP offering is the loss of revenue from the additional wireless services that we will not be able to offer. That is the opportunity cost of the project.

PROFITABILITY MEASURES

There are many measures of profitability that are used in business. These are applied to product lines and to specific products and services. We will discuss a few of these. Net present value (NPV) and payback period are concepts that might need to be considered for many telecom projects. These by no means constitute an exhaustive list of the profitability concepts that may be encountered in a project, but for most telecom projects, these will be sufficient.

Net present value is a concept that is used by financial analysts in ongoing business to determine the value, in today's dollars, of equipment that is owned and/or being considered for purchase. For a service provider, the engineers must select equipment to be used in the network, determine the amounts required initially and over time, work with purchasing departments to select and negotiate with suppliers, and provide documentation on this equipment to those departments that will install, use, and test it. Decisions on which products to select often involve the calculation of the present value of the outflows required for that equipment over a number of years. Project teams could well have to do such calculations to help justify, select, or evaluate the value of different equipment options that could be acquired for their projects. To determine the NPV of equipment or other assets, consider the cash flows associated with that equipment over a number of years. For telecom equipment, such analysis usually starts with the present (but it is possible to work backward from today as well to determine the values in earlier years).

The future value of a sum of money in today's dollars depends upon the interest or investment rate at which the money can be invested, and the number of years involved. A sum of money X in today's dollars, is $\$X$. Let us call that PV or present value.

Next year, if the money is invested at 10%, it will be worth PV plus 10% of PV. In two years, we would need to compound the interest, as the interest made in the first year would be included in the amount to be invested in the second year. Of course, this assumes that the investment or interest rate remains fixed over the period under consideration. In general, the future value, FV, of a sum PV is given by the equation

$$FV = PV(1+ i)^n$$

where i is the interest or investment rate and n is the number of years during which the money will be invested. Or, working in the other direction,

$$PV = FV/(1+ i)^n$$

If there is a series of sums to be spent in future years, then each of these needs to be brought back to the present value, and the overall total added, to determine the overall, or net, present value of the full investment:

$$NPV = \sum_{n=1}^{k} \{FVn/(1+ i)^n\} \text{ times the initial investment}$$

Often, engineers are called upon to calculate this value for technologies or other acquisitions for a project. It may be the case that two or more different alternatives are under consideration, and this value can be calculated for each, even though the costs would be different amounts spent at different times, giving one comparative measure of the options.

Payback period is a much simpler concept because it does not take into account the time value of money. To calculate payback period, one simply lists the expenditures by time period, such as year by year or month by month for the duration of the project or the lifetime of the product or service. The revenues during these periods are also listed, with all numbers being the actual value spent or made at the time. Amounts are not brought back to the present value. The costs can be subtracted from the revenues for each period to determine whether the product or service is overall positive or negative during each period. Then the amounts can be added together one period at a time until the running total hits zero. The time at which the sum of the net cash inflow and the net cash outflow is zero is called the payback period.

ESTIMATING THE COSTS

Project costs may be either direct costs or indirect, and these costs can be either capital or expense. Capital costs are costs of acquisitions, and tax law requires that these costs be treated differently on the books than costs that can be expensed. Capital costs result in the ownership of assets, most of which depreciate in value over time. Governments specify the depreciation rates that are to be used in accounting for these assets. Some assets, such as computers, which may cost lower amounts or depreciate to zero very quickly, can be expensed. Larger, more valuable assets must be capitalized. Both of these types of costs, if they will be incurred on a project, must be estimated at the beginning and included in the project budget.

Teams generally prepare the time estimates for the project, including the cost of manpower, any expense related to supporting the manpower, such as travel funding or meeting space and supplies, plus any required capital costs. They may include costs to pay for contracts or proposals that will be solicited. These costs must be estimated, as few will be firm before the project starts. Initially, the team will usually draw up a project budget, determining the cost of the major deliverables and the project as a whole. The costs for the time of the people will be estimated using the work breakdown structure as a basis. This structure is used to determine the project activities, and we need to assign someone to do the work on each of these. When the resource assignment is known, we can get a reasonable estimate of the time that will be required to complete each activity. The loaded resource rate of pay together with the time estimates lead to the actual estimate of the cost for the activity. This level of estimation is more granular than that used in many cases, but using this technique the team can ensure that nothing is missed and nothing is accounted for more than once. In the end, if the WBS is good, and the estimates of time and rates are good, this should provide the most accurate estimate of the costs. Since in the WBS the elements below any deliverable must completely describe the deliverable, once the costs of the bottom level elements are known, these can be added to give the higher level costs until eventually the cost of the full project can be determined. It is easily seen that the cost estimates then, at least for the cost of the work to be done, have the same problems as the time estimates. Some people will overestimate, some underestimate, and some will not have much confidence in the values provided. Every estimate should be shown with the confidence level for that estimate, to help the PM to determine just how much risk there is that the estimate might not be correct. The PM may include the accuracy of the estimates as one of the risks of the project to be monitored, controlled, and factored into the contingency equation. It is important that the PM learn as much as possible about the team members, so that he will be better equipped to understand which estimates might need to be adjusted to improve the accuracy.

Some of the costs will be determined by current prices for material or other purchased items. Some will be determined by bids. These exact prices might not be available until sometime during the actual project implementation, once the quotes have been received and negotiated. When this is the case, it will be necessary for the PM or the team to estimate these costs during the planning stages in order to confirm that the assigned budget is sufficient for the project work. The fact that the exact prices cannot be determined in the early stages is another of the risks for the project.

At this point, we should review the different types of estimates that exist. For telecom projects, as with most projects, corporate executives or the project sponsor have to make an early projection of the cost for the overall project. This is done before the decision to go ahead with the project, as an overall estimate of the project cost is needed to make the go-ahead decision. This much earlier estimate is quite different from the one described above, which uses the WBS to determine the cost of every individual activity and deliverable in the project. The main types of estimation used in telecom projects are:

- *Analogous estimates.* These are top-down estimates. These estimates generally apply to the full project, but sometimes analogous estimates are used to determine the cost of some major portions of some projects. These estimates are usually done by upper management, or by business assessment teams who will recommend which particular projects should move forward. They are based on costs of previous projects that produced similar deliverables, with some tweaking to take into account current prices and current situations that might differ from those for the past project. In many cases, they are very close to the actual costs that the teams incur when there is a history of similar projects. These estimates generally determine the project budget, or at least an initial project budget. It will be up to the PM to come back to the sponsor if the detailed plans show that this amount is not enough to effectively complete the project. For example, in the past, many companies estimated the project costs for adding features to a service such as a long distance service by comparing the development needed for the new service to that needed for previously developed features. The historical cost, with adjustments for differences between the new project and the previous one, in terms of changes in cost for equipment, changes in pay for people, and changes to the deliverables, produced a top-down estimate for the cost of the project to develop the new features.

- *Bottom-up estimates.* These are very detailed estimates taking into consideration all of the project deliverables and activities. They are formed in the manner described above, by assigning people's names to the activities in the WBS, then determining the amount of effort required from these people to complete the activities, and what charges the project will incur for this work. This determines the labor costs. To these costs we need to add any additional indirect costs, as well as the direct capital and/or material costs that come into play to complete the project. These should be the most accurate estimates of the overall project cost, but there is still risk that these estimates can be off. Sources of risk include variability in material costs, the experience level of those making the estimates of the time or costs, the number of elements in the project that are not completely clear, the management of the people doing the activities in the implementation phase, turnover in the project team (people leaving the company or getting sick, requiring that they be replaced by others who work at different speeds), and so on.

- *Parametric estimates.* These estimates are used much less often than the other two types, but for work such as programming or setting up test or call centers, this type of estimation might be useful. The estimates are made based on the number of repeating items that might happen in a project and the cost of each. A cost per line of code, for writing programs, would be an example of a parametric estimate, or, perhaps, the cost per station to equip a call center.

In some projects all of these types of estimates will be used. Usually, at least the first two types will appear at different times during the project.

PROJECT BUDGETS AND PLANNING THE PROJECT COST

All of the discussion in this chapter so far has covered planning and documenting the project costs. There are different ways to show the project cost and, generally, a project will use more than one of these. Three such ways are the project budget, the planned value curve for the cost expenditure, and the corresponding actual cost curve. All of these should include contingency. In the planned value, the contingency needs to be added into the ongoing costs, perhaps by including it in the costs of some of the activities. There are many ways contingency can be shown in the budget, and these are discussed below.

The project budget is one of the best known ways to show the project costs. The budget usually looks like an itemized list of the project deliverables. This budget is usually set before the project starts and, thus, it is frequently determined via analogous estimates. A typical format for the budget is shown in Figure 12-1. It generally lists the various high-level deliverables, broken down into subdeliverables or smaller items, with the overall total for the full project at the bottom.

Included in the costs, no matter what the format for the information, should always be contingency. Budgetary contingency is analogous to schedule contingency, as discussed earlier. The methodology for determining the amount of contingency to include is shown in Chapter 9. The more risky the project, the more contingency will be required to cover all the costs, as many things are likely to change from the

Merger of Wireless One and Two Connected Amalgamation of Network Inventory Databases	
Confirmation of Wireless One data	$ 20,000
Design mapping of Two Connected record structure to Wireless One structure	$10,000
Conversion of equipment inventory data	$25,000
Conversion of fiber optic cable inventory data	$25,000
Conversion of SONET inventory data	$25,000
Conversion of DS1/DS3 inventory data	$25,000
Conversion of site inventory data	$25,000
Cutover of Two Connected operations to new database	$ 3,000
Computers and servers	$ 6,000
User training and notification	$10,000
Contingency	$25,000
Total	$224,000

Figure 12-1.

initial plan and some of the impacts of the changes could be significant. The methodology calculates a single number for the total amount of contingency that should be included in the project budget. The question then is how this amount should be included in the project budget. The contingency amount calculated clearly does not cover the cost of all of the risks, or even a majority of the risks, that might materialize. It was calculated statistically, based on the likelihood that each risk might occur. It is expected that the entire amount will be needed in order to complete the project. The issue is only what the money will be used for, and this is something that the PM should manage closely.

There are a number of ways in which this could be incorporated into the budget, and some of these are more recommended than others. The amount could be included as a single amount, or broken into smaller portions that can be distributed through the budget. At one extreme, the full amount can be shown as a single entry in the budget. Handling it this way has the advantage that the PM can dip into this money as needed, giving him control over the way in which it is used. For an organization with a very good understanding of contingency, this would be a good way to manage this money. In the current telecom environment, with corporate management, corporate structure, and corporate direction changing frequently (and unpredictably), this method would not be recommended. With a change in management, or with the imposition of cost-cutting measures, which is the case in telecom today, it is highly likely that someone would see contingency funding, especially if it were large (as it is likely to be since the environment is risky) as a candidate for the axe. At the other extreme, contingency money could be spread evenly across all project activities. This is also not recommended, as each person working on each activity would see a budget that is a little more than needed, and happily spend the money, probably for creeping elegance to the scope. When a real risk hit, there would possibly not be enough left to cover it, and whatever money is left would have to be rescued by changing the budgets for upcoming activities. This would cause extra work for the PM and probably dissatisfaction amongst the team members. Therefore, a method of incorporating the funds that falls between these extremes is needed. The recommendation is that the total contingency amount be broken into smaller amounts, and that these amounts be distributed strategically through the project budget. The amounts should be associated with the deliverables, which are most likely to incur risk, so that there will be some amount of money immediately available should the risk materialize. There is often a correlation between schedule and cost contingencies: if a task is seriously underestimated, the extended duration will end up costing more as additional hours and days of effort are expended. The people undertaking these activities are likely to understand the concept of risk, since it is facing them directly, and should their risks not materialize, they should understand the value and need for saving the contingency portion of their funding so that it can be transferred to another activity for which the risk does materialize.

Contingency costs can be either added in with the costs for the work itself, showing one figure for each risky deliverable, or they can be shown with the deliverable cost as a separate entry for contingency. The separate entry method is better for overall control because it clearly delineates the contingency, making it easier to understand the true cost of the deliverable. However, if the environment is not mature enough in understanding of project management, as is all too often the case, the contingency funding could be at risk even in small amounts. In this case, it is preferable to allocate the contingency costs to those project costs at risk of overrun.

The second way in which the project costs are usually shown is with a curve that associates the expenditures with the time at which they will be made. This curve is called the planned value (PV), or the budgeted cost of the work scheduled (BCWS). The curve shown in Figure 12-2 is created by looking at the logic diagram for the project, which shows all the project activities lined up according to the time in which they will occur, as per the plan. For each activity, the full resource cost must be determined, including both the cost of the time of the person who will work on that activity, any overhead costs, any other indirect costs (i.e., the "loaded rate"), and also the expenditures for any items that will be purchased for the completion of that task. The planned expenditures are determined for each day or week of the project from the inception to the completion, and then these are added, time unit by time unit, to create a cumulative curve that shows the amount of money expected to be spent at any time from the beginning of the project to the completion. This curve is called the budgeted cost of work performed because the costs used to determine the amount to be spent by any given time are the amounts that were budgeted for the specific work. If one task is expected to be only partially completed by a given time, then only a portion of the budgeted cost, representing the portion expected to be incurred, would be plotted in the curve. Since these costs are the costs the team

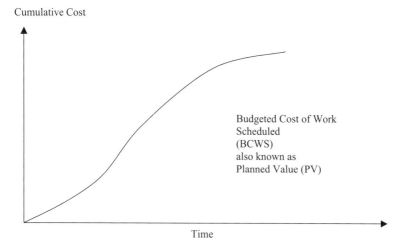

Cumulative Cost

Budgeted Cost of Work
Scheduled
(BCWS)
also known as
Planned Value (PV)

Time

Figure 12-2. Planned value curve.

plans to spend on the project, the curve is also known as planned value. This curve differs from the project budget because it considers costs at the activity level rather than the deliverable level and also because it links the costs to the time that the work will take.

Most projects do calculate the planned value information, and often this is plotted and controlled throughout the course of the project.

When plotting this curve, a question sometimes arises as to when to include a cost. For telecom projects, the project generally involves a significant amount of manpower and, in the past at least, most of this manpower was internal. People would charge their time to an account code on a weekly basis and, thus, the manpower costs increased gradually as the project progressed, with higher increases appearing when more people were charging their time to the project. However, many times outside contractors or consultants are used, for outside plant work, for management process work, and so on, and when these outside sources are employed the time at which the costs should be planned is not as clear. When a contract is signed or a bid accepted, there is already a commitment in place to pay for the work or the material. It is possible to allocate the planned expenditure at the time at which the contract is signed, since after that point the money has effectively been spent. Most project managers do not do this, however, because there is no transfer of money at this time, and if the material is not delivered or the work is not done as required, it is possible that the actual payment could be less than the initially planned amount. However, until there is any information about any changes to the costs, the amount to be included has to be the initially planned amount. Generally, people prefer to include this amount at the point at which the work will start, or even when the work finishes, with the expectation that the invoice for the work will follow shortly after completion. Even then, the accounting department may not record the expenditure until they have processed the invoice, which can be weeks later. No matter which point the PM selects as the appropriate point for showing the planned cost, there could well be a period of time between the appearance of the cost in the planned value and the appearance of the cost in the accounting records. When this happens, the PM is usually called on to explain the underexpenditure, which is not a true underexpenditure at all. The situation corrects itself once the accounting records catch up to the actual project status.

The third way that costs are shown is a recording of the actual costs, rather than the planned ones. This curve shows the actual cost of work performed (ACWP) or actual cost (AC). Ongoing actual expenditures are accumulated in a curve, which is generally plotted with the BCWS curve. This combined plot is intended to provide an instantaneous representation of the relative values of planned costs and actuals. People often try to use these to determine whether a project is over or under budget, but the curves must be scrutinized with great care for a valid understanding of the project expenditure status. An example is shown in Figure 12-3, and in this case one might say that the project is slightly under budget at the current time. In actual fact, it is not always possible to tell from the information shown whether the project is over or under budget. This is true for a few reasons, as we will discuss here. Consider the way these costs are determined. In the discussion above, we looked at including planned costs at different possible times based on contracts, work starting or finishing, and so on. In recording the actual costs, there should be consistency

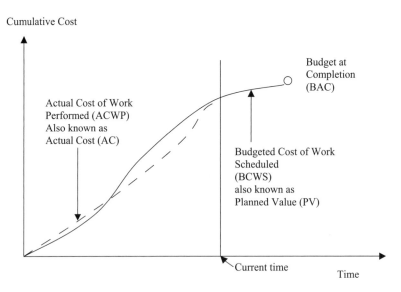

Figure 12-3. Actual cost of work performed curve.

with the planned costs. So if the planned costs were recorded at the time the work was planned to begin, the actual costs should be recorded at the time the work began, for consistency. If costs are incurred on the ACWP curve at the time of signing the contract, rather than when the work is actually done, this curve will trend too high. Conversely, if the costs are not included until the work is invoiced and paid for, the curve will underrepresent the actual costs.

Generally, direct comparison between these two curves is not a good comparison of expenditure versus budget at a moment in time. As the project continues, these variations are gradually averaged out, and if the project content is as planned, the curves will tend to converge.

BCWS is based on the work that is scheduled to complete by specific dates, whereas ACWP is determined by the work that was actually done by those dates. In most projects, particularly as the work progresses, changes are made in the timing or the sequence of tasks, and as a result the work performed often differs from the work scheduled. Thus, the two curves are measuring two different sets of work, and at any given point in the project the differences may be large. The measures used for the two curves also differ. The PV curve is using the budgeted costs to measure the cost, and the AC curve is using actual costs, which will vary with inevitable inaccuracies in the forecasting of the scope of the project tasks. For these reasons, a difference between the two curves does not necessarily provide good information on the project's performance level.

There is, however, another way to do a comparison between budgeted and actual expenditures, which requires no additional data than that needed to prepare these two curves. It is necessary only to repackage the information. The new curve is called the earned value (EV) or budgeted cost of work performed (BCWP). For this, we look at the work that has actually been completed at any given time, the same work that is considered in the AC curve. But instead of looking at the actual cost for each activity, we use the cost that was budgeted for the completed work. This then gives two sets of information about that same work—the planned cost and the actual cost. Comparing these two numbers is meaningful and this comparison does indicate whether the actual cost for that work is more or less than the planned cost for the same work. If BCWP − ACWP is greater than zero, then the value of the work performed is more than the amount spent for that work, which is good. If this number is negative, the project has spent more than planned on the work done to date, which is not so good. BCWP − ACWP is called the cost variance, or CV. We can also calculate a performance value called cost performance index, which is BCWP/ACWP. If CPI > 1, the performance is good; if < 1, the performance is not as good as it was expected to be. Although the BCWP curve eliminates inaccuracies caused by sequence shifts, it is still subject to error due to different methods of accruing project charges.

This same EV information can also be used to give one indication of the schedule status. If we compare the budgeted cost of the work performed to the budgeted cost of the work scheduled, we are using the same measure, budgeted cost, to compare two amounts of work. If BCWP > BCWS, then more work, or at least work that was deemed to be of more value, has been completed than planned. BCWP – BCWS is called the schedule variance, or SV. This is an indication that work is ahead of schedule, at least in one sense. We can also calculate a schedule performance index, SPI = BCWP/BCWS. If this index is >1, we are ahead of schedule; if it is <1 we are behind schedule. This third curve is shown in Figure 12-4. In the example shown in Figure 12-4, the project is both ahead of schedule and slightly under budget at the time under consideration. Following the curves throughout the project, though, we can see that this was not always the case. And it may or may not be expected to continue to be the case to the end of the project. In order to determine the future conditions, we would need to do an analysis of the situation. We need to determine why the project is ahead or behind where it should be in both budget and schedule. Then we need to assess whether or not this rate of progress will continue. If the rate is expected to continue, then the performance factors for the project to date can be applied to the budget and schedule so far to predict what the final cost and completion dates will be. However, if the future progress is expected to differ from that so far, new estimates must be developed in order to predict the completion values.

It should be noted that the specific values of the differences are not particularly important as the relative values can change often. What is important is the trend. If the PM sees a trend away from the budget or schedule, this needs to be investigated and corrective action should be put in place when necessary.

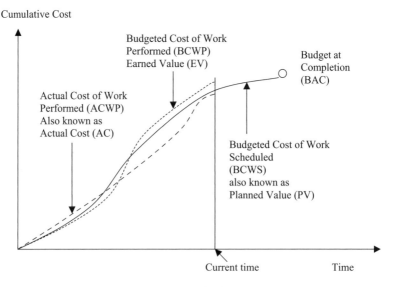

Figure 12-4. Earned value curve.

CHAPTER 13

MANAGING THE DEVELOPMENTS

In the *PMBOK,* fourth edition, page 43, we see the full set of process areas and the processes within these areas. The processes can also be viewed from a different perspective, classifying them with regard to the functions that they play in the phases of the project: initiating, planning, executing, monitoring/controlling, and closing. Table 13-1 shows both classifications.

Before the work can properly begin, the project manager must select the team members and inform them of the project background and how he wishes the team to operate. Once the work has been defined through input from the team members, they will create the schedule and check the workload on each of the team or other resources. If there are any imbalances or overloads, these need to be addressed. All of this has been discussed in previous chapters of the book.

Before the execution phase begins, the team should also define what to them constitutes project success, and how they will know when they have achieved it. If the project is to be managed properly, the team and the stakeholders must understand the importance of scope, time, cost, and risk control. They must use proper tools and techniques for this control. They will monitor each of these areas, watching for problems, identifying any variances from the plan, and planning corrective actions wherever needed.

As part of the planning, the team needs backup strategies for any foreseeable problems. They must decide which tools they will use to monitor and control the project work. In many cases, the company will already have standard tools that project teams are expected to use, but for new aspects the team may have to select or design such tools. Change management processes must be part of this to keep all the parameters in check.

The PM should ensure that every team member understands the importance of meeting the project parameters and goals.

Table 13-1. Project management process groups and knowledge areas

---

Initiating Process Group: Project Management Integration:
- Develop Project Charter

Project Communications Management
- Identify Stakeholders

Planning Process Group: Project Management Integration:
- Develop Project Management Plan

Project Scope Management:
- Collect Requirements
- Scope Definition
- Create WBS

Project Time Management:
- Activity Definition
- Activity Sequencing
- Activity Resource Estimating
- Activity Duration Estimating
- Schedule Development

Project Cost Management:
- Cost Estimating
- Cost Budgeting

Project Quality Management:
- Quality Planning

Project Human Resource Management:
- Human Resource Planning

Project Communications Management:
- Communications Planning

Project Risk Management:
- Risk Management Planning
- Risk Identification
- Qualitative Risk Analysis
- Quantitative Risk Analysis
- Risk Response Planning

Project Procurement Management:
- Plan Procurements

Executing Process Group: Project Management Integration:
- Direct and Manage Project Execution

Project Quality Management:
- Perform Quality Assurance

Project Human Resources Management:
- Acquire Project Team
- Develop Project Team
- Manage Project Team

Project Communications Management:
- Information Distribution
- Manage Stakeholder Expectations

Project Procurement Management:
- Conduct Procurements

Table 13-1. Project management process groups and knowledge areas

Monitoring and Controlling Process Group:
  Project Management Integration:
- Monitor and Control Project
- Integrated Change Control

  Project Scope Management:
- Scope Verification
- Scope Control

  Project Time Management:
- Schedule Control

  Project Cost Management:
- Cost Control

  Project Quality Management:
- Perform Quality Control

  Project Communications Management:
- Performance Reporting

  Project Risk Management:
- Risk Monitoring and Control

  Project Procurement Management:
- Procurement Administration

Closing Process Group:
Project Integration:
- Close Project or Phase

  Project Procurement Management:
- Close Procurements

Monitoring and control must continue throughout the project execution stage, with progress reporting systems in place and problem escalation processes defined to ensure that help can be available when needed to keep things on track or solve problems. Work results must be compared to the plans, and plans will need to be revised if they cannot be met.

The PM and the team should select and use a management and control system that has a strong emphasis on meeting commitments, that has well-defined policies that ensure that reporting is in place and accurate, and that provides visibility to senior management of the accomplishments of the team.

The PM should work with the sponsor and other senior management to ensure that they will provide any assistance needed. They can assist with coordination efforts, help to provide a good home for the project, and allow the team to adapt to change when this is needed. The senior management should provide the PM with information about corporate strategic plans, and work with the PM to build project plans that are consistent with the corporate plans. Management should always demonstrate visible support for the project, and PM should ensure that this happens.

Management and control will be easier for all concerned when the project (time, cost, scope, and quality) specifications are clear and realistic. Many telecom projects, such as customer network designs or the development of customized applications, are done in direct support of a client. These projects cannot complete successfully without support from the client organization. Client organization support also needs to be managed. They should be interactively involved in the establishment of goals and specifications, in the coordination efforts, in the provision of information about their requirements and the performance of their existing applications and networks, in providing access to their site or other resources as needed, and, possibly, with some procedures if these overlap with client operations.

The PM must create and maintain a team atmosphere that is professional, with focus on achieving the project goals, objective communications, working together with team members and the client, possibly competing within the team to obtain better results, clear role definition for all activities, approvals and decisions, quality work and quality products, and in some cases determining appropriate conditions for project termination. The PM should research and identify best practices which apply to the project and its environment, and make sure that these are used. The concept of best practices as methods by which the company can attain a competitive advantage is generally understood in telecom companies, and these can be applied to projects.

From time to time as the project moves forward, the team must review the objectives of the project and analyze the environment within the project, within the company, and outside the company to ensure that the objectives are still appropriate, and that the current progress will allow these to be met. If problems accrue, these objectives and/or the plans may need to be changed. This should be done only after serious consideration of the new information by the project team, and with the knowledge and concurrence of the client, management, and key stakeholders.

The PM must also be alert for potential conflicts, as once the work begins in earnest these are very likely to arise. Many factors, generally related to change, can contribute to conflicts, so the team should be alert in watching for these. Such factors include:

- Possibility of project cancellation or direction change
- Errors or failures of the team or client to meet some expectation
- Schedule commitments that are problematic for the team
- Poor control of project change requests or design changes
- Poor project cost accounting
- Change in project leadership or key team members
- Failure to receive a critical input or information
- Failure to receive anticipated approvals

It may be necessary to consult with the sponsor about management's view of the project and the environment to obtain new or different information that can help reduce conflict, stimulate teamwork, or improve project direction. A positive assessment from upper management or an understanding of issues or changes that the team must deal with can have very positive impact on a project team.

In the implementation stages, the PM should also maintain communication with the functional managers of the team members, regardless of the organization structure used for the project, as functional managers are also stakeholders in the project. They may be able to help with issues or problems in some cases, and their people can feel more secure knowing that their usual management understands what they are working on.

The PM, or a designated person on the team, must closely monitor the work packages, the schedule, the budget, the risks, and the quality. Periodically, someone should review the status of each project work package to obtain a clear understanding of what to expect going forward as well as looking back. Key items to monitor include:

- Time to complete
- Cost to complete
- Work to complete
- Potential risks in the near term
- Potential risks in the longer term

When problems arise, the team needs to consider the issue behind the problem and possible solutions. The team should prepare a list of alternatives and assess each for the value it brings, at what cost. If an option requires a delay of the project completion date or a key milestone or deliverable, or an increase in cost of the project, the team must consider the impact of this delay or cost on the project and the stakeholders. Questions that could be asked include:

- Is a time delay acceptable to the customer? The customer may be internal or external, but in any case there is likely to be some impact, and this must be managed.
- Will the time delay change the completion date for other projects? Are the stakeholders of the other projects agreeable to a change to their project timelines?
- What is the cause for the time delay or cost overrun? How can this cause best be handled?
- Can resources be recommitted to meet the new schedule?
- What will be the cost for the new specific alternative?
- Will implementing this alternative impact the company's ability to procure future contracts?

- What can be done to reduce the remaining costs once this alternative is implemented?
- Who will cover the cost overrun?
- Can the project requirements be renegotiated to stay within cost or schedule?
- Are the time allocations and budgeted costs for the remainder of the project accurate?
- Are the specifications negotiable?
- What are the advantages to the company and customer for specification changes?
- What are the disadvantages to the company and customer for performance changes?
- Will there be a product or employee liability incurred?
- Is there any union impact?

The team must analyze all the feasible alternatives and decide how to proceed. There are many processes that can be used for this. Some of these are quite informal, just involving discussion amongst the people involved. But for most telecom projects, the budget and impact are considerable, so data should be carefully collected to support all predications and decisions, and the best possible tools should be used to ensure that the decision is optimal for the project in its environment. The team may decide to use a decision tree model with costs, work objectives, and schedules, to estimate the probability of success for each condition leading from the decision point.

Corrective actions can include such things as changing or easing the specifications; reassigning the work; working longer hours, with or without pay; changing vendors; using different equipment or configurations of software and/or equipment; bringing in outside contractors, which today may be former employees who can arrive with much of the required background; and proceeding prior to or without testing or approvals.

Most of the earned-value concepts were described earlier, but here we add four additional concepts that can be added to the previous material for use in managing the developments as the project proceeds. The terms not previously discussed are estimate at completion (EAC), estimate to complete (ETC), schedule performance index (SPI), and cost performance index (CPI). EAC is a concept that is needed if the performance is not exactly on track, and it usually refers to the cost of the project. If the budget is not tracking, then the team needs to determine whether or not the project budget will be met. They need to provide a new estimate for the cost of the project, based on the amount spent to date, the specific work that has been done, and the expectations for the remainder of the project. If the performance levels that have occurred to date are expected to continue, then the team can apply formulas to the amount spent to determine what the expected overall expenditure will be at the end of the project. For example, if the project has been running over budget by 20% at the examination point, and the performance levels are expected to continue, then the BAC will not be met and the EAC will be 20% higher than the BAC. However, if the performance levels are not expected to continue, then the estimates for the remaining work must be determined, based on what this work is and what the team now knows about the work. These estimates are determined by processes similar to those used for the initial estimates. In both cases, the team calculates an estimation of the expected project cost at the end of the project. The ETC is the difference between the current amount spent and the EAC. SPI and CPI are both performance indices showing how well the team is doing in comparison to the plan. SPI = BCWP/BCWS and CPI = BCWP/ACWP.

The earned-value concepts should be part of this overall management, showing the team the trends before things get too far off track and allowing for decisions to be made based on proper analysis and comparisons. The team should prepare regular performance reports including BCWS, BCWP, and ACWP information, with predictions for requirements going forward. The report might also include information on equipment or software procurement, delivery, and usage. The team should provide (and probably post on the intranet) regular status reports that describe the status of the work, and include information on variances such as SV and CV. The future predictions need to be clearly thought through, with consideration for why the work is currently where it is, and whether or not current rates of progress are expected to continue. The earned-value analysis parameters of EAC, ETC, SPI, and CPI should be calculated, but the team must also understand that calculating information is not the same as control. They must understand the reasons that the numbers are as they are, and then develop forward-looking projections based on the best estimates of what is to be expected going forward. If the future progress does not meet project parameters, the actions mentioned above must take effect to try to return the project to the (revised) desired track. In addition, the team should prepare and share exception reports identifying exceptions, problems, or situations that exceed the threshold limits on such items as variances, cash flow, resource assigned, and other such topics. These also need to be addressed with good decisions based on facts, to allow the project to meet requirements and expectations.

CHAPTER 14

MANAGING THE PEOPLE

In the telecom environment, projects are a way of life. Most of the work performed by a telecommunications company is day-to-day management and running of the business, but in order to ensure that the products, services, and support systems continue to be state of the art, many projects of varying types must be undertaken within every company. We have noted earlier that there are many different types of companies, often offering the same sort of services or products in this now highly competitive environment. It is not surprising that projects that occur in one company also occur in others, but the projects look very different when they occur in different types of companies. For example, an incumbent telco, a cable company, and a new entrant to telecom might all decide to offer a wireless multimedia service within similar timeframes. The telco will build the service on an existing network as much as possible with already installed technologies and operations systems, using staff that has a good understanding of how to develop and build telecom and probably also data services. But they might be newcomers to some of the multimedia areas, perhaps with some basic training and a little experience building one or two similar services. The cable company will also build the service on an existing network, but it will be one that is quite different in technology, architecture, and functionality from the telco network. They will also use experienced staff, but the staff will be experienced in developing broadcast services. Thus, the backgrounds of the people will be different. Both will require people with knowledge of engineering; marketing; perhaps sales; the multimedia technologies under consideration; business development; operations, including ordering, testing, monitoring, and billing; and likely a number of other areas such as programming or purchasing. Both companies will already have many people with background in most of these areas.

In the case of the start-up, the end results of the project will be very similar to those required by the other two companies. However, the company probably does not already have an existing network, so they will have more freedom in selecting the technologies, but less already available support. This allows them more freedom in the design of the service because they are not constrained by the current technologies and capabilities. If the company is to succeed, the people working for the company will need to include some who have significant experience already in the industry, and who bring different backgrounds and skills to the table. Assuming the start-up was created to introduce this new service, it will probably have a higher percentage of people who do not have industry background but who have a solid knowledge of the newer technologies, possibly stronger than that of the people in the incumbent companies. Whereas the people in the established companies will understand and appreciate (either positively or negatively in some cases) many established practices and procedures for working and for building new services, those working for the start-up are likely to be less encumbered by corporate overhead, possibly allowing for more innovation in the design. Thus, although all are designing and building similar services, the teams, environments, and products can look quite different from each other. All need to use project management because timely completion of the project, with a high-quality service at the best possible cost, is critical to each, for different reasons. Although the project plans of these three typical companies may look quite different, the PM processes and techniques that are advisable for one company are essentially the same as those that will prove valuable to the others.

In addition to the development of new services or products, many telecom projects involve developing new processes, cultures, or relationships, given the rapidly evolving nature of the business and the environment. The information in this chapter applies equally in these projects.

ORGANIZATION STRUCTURES FOR PROJECTS

As in companies in other business areas, telecom projects are generally overlaid on top of an operating business. There are various ways in which this can be done. We will look at five such structures here. In every case there is a company structure already in place, and in that structure there are people working on all aspects of the business. Usually, this is done within functional groups, but sometimes the business is split along lines of business at the highest level, with functional organizations created within each business line. The projects must then fit into this existing structure, usually using people directly from the existing staff, possibly augmented by one or a few outside contracted people hired to provide something specific to the project. For the duration of the project, it is necessary that these people work together as a project team, even though it is highly likely that they come from different disciplines and possibly different business areas, and they may not know each other at the beginning of the project. Thus, there is a requirement that the person managing the project find ways to ensure that the people will work together harmoniously. What it might take to do this will differ from one project organization structure to another. The five different structures that we will consider are projectized, strong matrix, balanced matrix, weak matrix, and functional.

Probably the first structure that comes to mind for most people is the projectized structure. In this structure, the sponsor hires someone to act as the full-time project manager for the project, and this person recruits, from whatever source is appropriate, full-time people who fulfill the various roles in the project. The team should consist of people with all the skills needed to do the project work. Since the people are employed full time on the project, reporting to the project manager, it is quite possible that some people may need to cover portions of the project that do not fall within their usual primary responsibility areas. It is also possible that at any given time during the project some people are not needed full time, and during these times the team may decide to share work that might normally go to others. Doing this can keep all team members occupied, which uses the resources more effectively than having people idle. But it might also result in having some activities take longer than they might if there were an expert in the area handling the activity. This structure is excellent for building the togetherness of the team because they all work for the same boss, and they might well all sit in the same area. It works well even if the team is not colocated because people tend to communicate with others who also work for their boss. It provides good focus on the project and the objectives of the project because everyone works full time on this project, creating a natural interest in project-related information and issues. In addition, project-related communications between different functional team members occur easily since the people are all working in the same group.

This structure is often used by service providers for the development of a new service and by manufacturers for the development of a new product, when the company wants to have the team working together for the duration, constantly throughout the project, with a great deal of focus on the project. It sits are one end of the continuum when considering project structures. At the other end of this continuum is the functional structure. Let us look at it next.

In the functional structure, all team members continue to work for their usual supervisors and, in most cases, they continue to work on their usual job functions as well as doing the project work. There is no one with the title of project manager or project coordinator, but it is usual to find someone designated as the person who will keep the project work moving forward. If most of the team members belong to one functional group, then the manager of that group might well take the lead in ensuring that the project work is done. If one group needs the project results, the manager of that group might take the lead or ask one of his people to do so. Thus, if the project aims to develop new trouble-handling processes, the people involved might come mainly from operations, but perhaps engineering and sales as well. Operations is very likely to take the lead. In this structure, people can work on the project part-time and, in fact, team members might participate in a regular job plus multiple project teams at the same time. This can provide the team with some very solid functional background, as each resource is using the same sort of skills needed for his or her regular job and, in addition, each has easy access to the guidance of his or her regular supervisor plus colleagues with similar backgrounds. People can work on projects and not worry about not having their regular jobs to return to at the end of the project, since they continue to report to their regular bosses. If the team has members from various functional groups, this can make communications difficult, as these people may not be in the habit of communicating directly with each other in the course of day-to-day business and, in some cases, especially if there are disagreements or issues, both of the department managers might also have to become involved in communications. The focus on the project is frequently low for most team members because they are also expected to continue their usual work. On the other hand, resources are used quite efficiently since people who are not needed full time for their own portion of a project will continue to use any remaining time on their own work or other projects.

Between these two extremes there are three additional ways to set up a structure for a project, all three classified as matrix structures. The matrix structure attempts to gain the advantages of each of the first two structures and to minimize any problems. In the matrix structure, there is a project manager for the project, or at least someone designated to coordinate the project work. The people on the project team do report to this person, generally in a dotted-line reporting scheme because they also continue to report to their regular bosses as well. Thus, every project team member has two supervisors—the usual functional manager and the temporary project manager. Since there are usually multiple functional managers involved, and these will have differing objectives and differing views of the project and other priorities, so there will no doubt be conflicts from time to time in their expectations of the team members. This places each of the team members in a difficult position and can cause problems in getting the work done. The PM has to deal with these problems as they occur.

There are three different matrix structures, called strong matrix, balanced matrix, and weak matrix. The structure for each of these is the same, but in a strong matrix structure the company or organization places a strong focus on the project rather than on the day-to-day functional work. Thus, this structure has many of the attributes of the attributes of the projectized structure. In the balanced matrix structure, the company places equal weight on the project and the regular work. In this case, the functional manager and the project manager share the management responsibilities for the team members. This is usually done by allowing the functional manager to decide who will work on the project and how the work will be done, while the PM decides what work is required and when it will be done. Finally, the weak matrix is a structure in which the organization places more emphasis on the ongoing work than on the project. In this structure, the PM is usually just a project coordinator, which places the person in an even more difficult position, having to convince people to do the project work when others potentially see this work as not very important relative to the ongoing work load. Obviously, in this structure the PM needs to have very good influencing skills to get the work done properly and on time.

Some form of matrix approach is the most common environment for projects. The strength of the matrix structure is generally defined by the corporate culture, but there are sometimes cases where different departments participating in the project will have different ideas about the relative importance of project and ongoing work; this can be extremely challenging to the PM.

MANAGEMENT STYLES

Many of the areas that the PM must understand and consider are management techniques or skills that apply just as much in general management as they do in projects. But since they do apply in project management, we should at least mention them and maybe talk a little about them. Topics covered range far beyond those covered here, to include such things as:

- Creativity
- Problem solving
- Innovation
- Change management
- Strategic thinking

and many more.

Project managers are managers and, as this book has hopefully shown, project managers require knowledge of and skill in many areas of traditional management as well as many areas that are specific to projects. The style of management is one thing to be considered.

When it comes to dealing with people, there are many styles of management that are found in organizations. There is much literature available on the pros and cons of various management styles, with different classifications of styles defined by different researchers, showing the different situations in which one style or another is most effective. Each manager has a style of management, and the team members quickly learn what this style is, often without being aware that they are doing this. Armed with an impression of the PM's management style, team members alter their own methods of interaction to maximize their own advantage in the workplace. Each team member will form preconceived ideas of how the PM and other key managers will react under specific conditions and they will interact with these managers in a manner that they deem to be the most appropriate to accomplish their own interactional goals. It is helpful if the PM can be aware that this will happen, and perhaps monitor his reactions and behavior to encourage the types of behavior in others that are needed to make the project successful. Many telcos and manufacturing firms provide training in management for their managers to enable the company to improve the way in which they operate.

In this chapter, we will just touch on some of the concepts considered in discussing management styles. Management styles are also related to personality types, so some of the literature talks about personality types as well. For more detailed information about personality types, reference material by authors such as Myers and Briggs is very interesting. They map personalities along two scales, representing rational or judging functions (thinking), and feeling and irrational or perceiving functions (sensing and intuition). They classify people into one of 16 different personality profile types depending on how they use these functions. People who fall into the different personality categories mapped onto these axes are likely to share management styles with others in the same personality category. Taking this one step further, those with similar management styles will be expected to react to different project situations or conflicts in similar ways. Therefore, it is interesting and helpful to understand the personality styles and management styles at least a little.

Many authors write about management styles and management techniques and their effects on people and productivity. Some more prominent authors in this area include Edward de Bono, Robert Heller, Edward Deming, Robert Tannenbaum, and Warren Schmidt, but there are hundreds more to select from. A person's management style reflects the way in which that person leads or manages people. Each expert has a set of classifications of types of styles, and while there is similarity from one to another, they are not all the same. In one view, styles range from being very autocratic to being very permissive. Autocrats take more personal control of situations, people, and directions, whereas more permissive managers expect and require team members to not only handle their own work, including decisions that are within their own realms, but also to participate in the decision making for the project as a whole.

The following discussion taken from Wikipedia* explains one view of the management styles, showing some of the pros and cons of each. Others have somewhat different classifications, but this one is illustrative of a fairly typical description of styles. A few editorial notes have been added to the Wikipedia quotes here and there.

Autocratic

An autocratic or authoritarian manager makes all the decisions, keeping the information and decision making among the senior management. Objectives and tasks are set and the workforce is expected to do exactly as required. The communication involved with this method is mainly downward, from the leader to the subordinate. Critics such as Elton Mayo have argued that this method can lead to a decrease in motivation from the employee's point of view. The main advantage of this style is

*Wikipedia, http://en.wikipedia.org/wiki/management_styles. Accessed June 2010.

that the direction of the business will remain constant and the decisions will all be similar. This, in turn, can project an image of a confident, well-managed business. On the other hand, subordinates may become highly dependent upon the leaders and supervision may be needed.

Paternalistic

A more paternalistic form is also essentially dictatorial, however the decisions tend to be in the best interests of the employees rather than the business. (Note: It should be pointed out that the decisions reflect the *manager's* perception of the best interests of the employees, who are unlikely to be consulted!) A good example of this would be David Brent running the business in the fictional UK television show *The Office* (the U.S. version office manager is Michael Scott). The leader explains most decisions to the employees and ensures that their social and leisure needs are always met. This can help balance out the lack of worker motivation caused by an autocratic management style. Feedback is again generally downward; however, feedback to the management will occur in order for the employees to be kept happy. Assuming that the manager's judgment is correct, this style can be advantageous and can engender loyalty from the employees, leading to a lower labor turnover, thanks to the emphasis on social needs. It shares similar disadvantages with an authoritarian style, as employees become highly dependent on the leader, and if the wrong decisions are made, then all employees may become dissatisfied with the leader.

Democratic

In a democratic style, the manager allows the employees to take part in decision making; therefore, everything is agreed to by the majority. The communication is extensive in both directions (from subordinates to leaders and vice versa). This style can be particularly useful when complex decisions need to be made that require a range of specialist skills; for example, when a new ICT system needs to be put in place and the upper management of the business is computer illiterate. From the overall business's point of view, job satisfaction and quality of work will improve. However, the decision-making process is severely slowed down, and the need of a consensus may avoid taking the "best" decision for the business. It can go against a better choice of action.

Laissez-faire

Chris Fest is the founder of this unique management style. In a laissez-faire leadership style, the leader's role is peripheral and staff manage their own areas of the business; the leader, therefore, evades the duties of management and uncoordinated delegation occurs. The communication in this style is horizontal, meaning that it is equal in both directions; however, very little communication occurs in comparison with other styles. The style brings out the best in highly professional and creative groups of employees; however, in many cases it is not deliberate and is simply a result of poor management. This leads to a lack of staff focus and sense of direction, which, in turn, leads to much dissatisfaction and a poor company image.

Styles of encouraging people to do the work can also fall into quadrants. One such breakdown, found on Wikipedia,* which looks at the different ways in which a manager can convey a request to employees, lists four ways to give work assignments, with comments on each:

1. *Telling.* This works best when employees are neither willing nor able to do the job (high need of support and high need of guidance).
2. *Delegating.* This works best when the employees are willing to do the job and know how to go about it (low need of support and low need of guidance).
3. *Participating.* This works best when employees have the ability to do the job, but need a high amount of support (low need of guidance but high need of support).
4. *Selling.* This works best when employees are willing to do the job, but do not know how to do it (low need of support but high need of guidance).

Some managers seem to need to remain very close to the work and the employees, unable to let go of the control over the way in which the work will be done. Such micromanagement is generally received quite badly by the team members, and using this technique can backfire, producing behavior that works against the goals of the manager. Other managers are more hands-off, and this style also can create problems if closer guidance or supervision would be appreciated by the team members.

There is no always "right" way to manage an organization, merely ways that work better than others in specific situations, influenced by corporate and departmental cultures, the abilities of the individuals in the organization, and, often, the characteristics of certain influential individuals within the group. Project managers should assess the project, the requirements, the team, the environment, and their own personalities to consciously decide and implement management techniques that are most appropriate, taking into account all of those.

LEADERSHIP

In order for projects to complete successfully, leadership, in addition to management, is needed. The PM needs to be a leader in addition to being a manager, and there are places for others to take leadership roles within project teams as well.

*http://wikipedia.org/wiki/Situational_leadership_theory. Accessed June 2010.

These other roles could be technical leadership, strategic leadership, or leadership of the team toward the goals of the project.

What is leadership? There are definitions in dictionaries, which mainly amount to stating that a leader is someone who is followed. Therefore, anyone who can get others to follow his guidance or direction is a leader. There are many ways to enable someone to get others to follow. These can be divided into essentially two categories: use of authority and use of influence. Use of authority occurs when the leader has been granted authority over others. When this happens, most others will follow, barring some extreme situation that might cause them to not do so. Thus, having the title and rank of manager brings with it benefits, because those in positions of lesser authority will do as the manager requests simply because of his title. For the most part, people have demonstrated management if not leadership skills in previous positions without the title, in order to be granted the management title. So, in addition to the title, most managers will use their leadership skills to get the work done properly. This means that the leader has to exert influence. There are many sources of influence that can be used, both positive and negative. Influence can come from technical strength or other relevant experience, which allows the technical experts to gain the respect and following of their peers. This is frequently a factor for leadership within telecom teams since people are working with such a variety of highly complex technologies, with more appearing almost weekly. Having a personal relationship with someone in authority can cause others to follow you, with generally mixed results, as such a relationship does not necessarily equate to having skills useful in getting the job done. Being strong enough either physically or in character to use force or threats is one of the negative ways to use influence. This often amounts to bullying, and though there are a few situations in which this might be useful (e.g., someone is in danger and has no other way out of the dangerous position, maybe not enough time to use other forms of influence), this is generally not accepted as a good technique. Humor works and personal charisma can be a very effective tool. These must be consciously used to assist in getting the work done, rather than being seen as a replacement for working.

Effective leaders have the ability, using whatever techniques work for them, to line up others behind their objectives and get the others to move toward achieving these. Some people believe that leaders are born or, in other words, that some people possess leadership traits whereas some do not; others believe that leadership can be developed. Again, there is quite a lot of literature discussing this, and it is well worth the time for a PM to review and consider some of this. The PM needs to hone the leadership skills that he has, and watch for leadership in his team members. Positive reinforcement for showing leadership can help to encourage people to use these skills to the benefit of the project.

For more information on leadership, it is useful to review the models in the literature. One reference could be the management styles model by Bonoma and Slevin, which divides leadership into four categories—shareholder, consensus manager, consultative autocrat and autocrat—and describes the characteristics of each.

TEAM BUILDING

One person's idea of reality is just a bit different from everyone else's. Every organization and, often, different departments within the same organization, have different attitudes, and ways of dealing with challenges. The project manager has the responsibility for helping the team made up of very different people to act and work as a team rather than as a group of individuals. Project objectives are much more easily and effectively met by people functioning as a team than by many individuals, regardless of the project type or the skills involved. For the team to be effective, the required skills have to be provided by the team members, so the first responsibility is selecting the right set of people for the jobs at hand. The people should be selected for the skills that they possess to enable the project work to be completed, but consideration must also be given to how each individual will interact with the others. Someone with strong technical skills but who cannot interact well with others is generally not a good choice.

Given that projects are completed by teams, and that these teams frequently consist of people who do not usually work together, building and maintaining team spirit are very important skills for the PM. This is another topic that has high importance in general management. There are literally hundreds of authors who have published suggestions for team building, with courses available in every major city.

Referring again to Wikipedia,* we find the following information about team building:

> The term *team building* generally refers to the selection, development, and collective motivation of result-oriented teams. Team building is pursued via a variety of practices, such as group self-assessment and group-dynamic games, and generally sits within the theory and practice of organizational development.

When a team in an organizational development context embarks upon a process of self-assessment in order to gauge its own effectiveness and thereby improve performance, it can be argued that it is engaging in team building, although this may be considered a narrow definition.

The process of team building includes:

- Clarifying the goal and building ownership across the team
- Identifying the impediments to teamwork and removing or overcoming them, or if they cannot be removed, mitigating their negative effect on the team

To assess itself, a team seeks feedback to find out both its current strengths as a team and its current weaknesses versus the needs of the project.

*http://en.wikipedia.org/wiki/Team_building. Accessed June 2010.

To improve its current performance, a team uses the feedback from the team assessment in order to identify any gap between the desired state and the actual state, and design a gap-closure strategy.

According to the PMBOK, team development is one of the PM processes, for which the tools and techniques that should be used include:

1. Team-building activities
2. General management skills
3. Reward and recognition systems
4. Collocation
5. Training

In this book, we only touch on these topics. There are many books on team building activities, so PMs can find many to choose from, and doing some research in this area is well worth the time. Some of the techniques for team building involve holding a kickoff meeting and ensuring that everyone is well informed about the project goals, directions, and expectations. This is always important. Other suggestions include maintaining a professional attitude, ensuring that all team members have complete project information at all times, and treating everyone fairly and professionally. Of course, each of these is also important, and should always be done on every project. Yet other suggestions include sharing of material with project identification, such as caps and t-shirts; holding shared social events such as lunches, dinners, and parties; attending shows; or even spending weekends off-site playing trust-building games. It is important to select references that are appropriate for the members of the team, for the project to be undertaken, and in which team members will be able to participate. Something that requires a weekend away might be fun and interesting for some groups, but may not be possible for, say, a single parent on a limited budget with young children. Not all techniques are appropriate for every project. The project manager must work with the team, the environment, the budget, and the requirements to select appropriate team-building methods.

Most project managers already have some level of general management skills before they are selected for this position. However, everyone can benefit from improving management skills. Again, these are not project-management specific, and much information is available from published management sources.

The rewards and recognition that might be appropriate should be consistent with the motivation theory and techniques the PM selects. Motivation theories such as those published years ago by Hertzberg or Maslow are still popular theories. These specify the sort of things that should motivate team members, as well as what is not likely to motivate. The PM can review motivation theories and determine which he wants to apply. This will help to define the sort of things that should motivate the team, and the PM can then ensure that he will be able to provide such rewards and recognition. At the professional level, any form of true recognition is a motivator, and every PM should always provide recognition freely to team members at any opportunity.

In the past, having teams that are not colocated was not recommended. In telecom today, many teams are not colocated, but with the current Web-based and phone technologies it is possible for people even in different countries to work together in real time quite effectively. IEEE teams organizing conferences and building certification programs are two such examples. Even in these cases, though, it is important to recognize the togetherness that is lost when teams are not colocated and to manage around this. Managing teams that are not colocated does take more work and focus than managing teams that have all team members located in a single location.

It is often a good idea to incorporate team-building activities into the project directly as much as possible. Creating the WBS and the logic diagram as a team are good examples of items that are directly project related and which can build the team dynamics as everyone works together to create these.

A professional atmosphere, a challenging project, a "place" for the team to meet, even if it is Web-based, will all help to strengthen the feeling of togetherness that teams need to be effective.

MOTIVATION

The responsibility for motivating the team members to do the assigned work in the best way possible and to watch for ways in which project problems can be avoided or solved without constant direct supervision belongs to the PM. Motivating individuals requires some knowledge of the people and what motivates each of them. Some of the motivators are individual; what motivates one person will not necessarily motivate someone else. But there are some things that fit for most people within a culture, whether the culture be a nationality, a regional culture, or even a corporate culture. Some cultural motivators can be easy to identify, for instance, whether people are conscious of status or make displays of material wealth. But many things are unspoken within any culture. Generally, rights are assumed, values are implied, and needs are unspoken. Thus, the PM needs to interact with the team members and pay close attention to things they reveal about themselves in order to determine what might motivate each person.

There are many different theories of what motivates people. It is wise for a PM to be aware of some of these, to help in finding ways to motivate different team members. Many of these theories have been around for decades. The theories look at people and their interactions from a number of different perspectives.

One of the best-known theories of motivation is attributed to Herzberg. Herzberg divides factors into two classes: motivators and hygiene factors. The factors he called motivators are things that are inherent in the job itself, such as responsibility or work that is satisfying to the individual. Giving more of these factors will motivate the person to work harder or do better work. Hygiene factors, on the other hand, are more basic needs, and are related to the environment more so than the actual work. Improving hygiene factors will not usually motivate the person; these are considered as basic requirements for an acceptable work environment, but if a person does not receive enough of these things, he is likely to become quickly demotivated. Items in this category include salary, basic work environment issues such as space and temperature, work hours, individual status within the organization, or extra benefits. There is not really a clear delineation of such factors; what one person may consider "table stakes" may be considered a significant perq by someone else. However these are characterized, these factors by themselves do not provide a motivated enthusiastic employee. A PM needs to strive to make each person's job as interesting as possible for the individual, with inducements beyond those which the person will consider a basic minimum. A salary increase is not likely to motivate better work; dissatisfaction will occur, though, if a person feels that the salary he is being paid is below what his work is worth. This theory is known as a "need-based" theory.

Another way of looking at this is presented by Maslow's Hierarchy of Needs. Maslow stated that people have needs, and they can be motivated by things that satisfy these needs. However, the needs occur in a five level hierarchy, and those at the bottom need to be satisfied first. Once the bottom-level needs have been satisfied, the person's attention will shift to the next level. Significant motivators start to appear in the higher layers, so to motivate someone effectively the PM needs to ensure that the lower level needs are met before addressing those at the higher levels. The hierarchy is as follows, with the highest level needs appearing first, followed by those at the next level down the scale:

1. Self-actualization (creativity, problem solving)
2. Esteem (self-esteem, confidence, achievement)
3. Social (friendship, family)
4. Safety (security of family, of self, of employment)
5. Physical (food, water, sleep)

These particular theories are not new. Maslow first published his theory in 1943. These two theories really present the same basic idea. Herzberg's "hygiene" factors are roughly equivalent to the bottom two layers of the Maslow model, and his "motivator" factors to the top three. The message here is that the best motivation arises from the team members' ability to feel creative and having accomplished something of importance, but the PM should not expect to see this happen if the team members are occupied by issues of basic work environment.

CONFLICT

According to Wikipedia, conflict is a state of discord caused by the actual or perceived opposition of needs, values, and interests. A conflict can be internal (within oneself) or external (between two or more individuals). Conflict as a concept can help explain many aspects of social life such as social disagreement, conflicts of interests, and fights between individuals, groups, or organizations.

Conflict is usually regarded as a purely negative concept, but to assume that this is always the case is shortsighted. Issues that occur during the course of a project are often understandable in terms of conflict, leading to a better analysis and driving the necessary decisions to resolve the issues. A task that has slipped due to any number of project factors will create a degree of conflict if a group of people are now required to take on a task at the same time when they have another scheduled. This situation is appropriately called a "resource conflict." In many cases, it is just not possible to get the two tasks done, and the conflict raises the issue in the attention of the PM and other managers who may be able to deal with the resource issue. The fact of conflict focuses the attention of the people who need to make the appropriate decisions to resolve the problem. The success factor here is to address the conflict situation to the satisfaction of the people involved, before they start viewing the situation as a personal affront.

Conflict can occur in almost any situation but is more likely in a project environment because every project brings change, and most occur in a setting which itself is a change from the usual work environment. Thus, any project team should expect to encounter some conflict, and the PM needs to be ready to handle it. Of course, the telecom environment is one that is experiencing rapid change already, even without projects interfering, so we should expect to find many conflicts in every project.

Given that teams are multidisciplinary, with different objectives, backgrounds, and perspectives, it is to be expected that there will be times of disagreement, in which strongly held honest opinions may lead in very different directions. People may misinterpret the motives of others from different departments or companies. One group may assume that they are providing information about some aspect of a project, whereas others might see this as a request for a different direction or methodology. The second group might feel pressured, whereas the first group can be confused about their reaction. If this confusion cannot be resolved, bad feelings could occur and much time can be lost in completing the project work.

The responsibility of the PM in such situations is to resolve the conflict so that people can return their focus to the work at hand. Unresolved conflict, as well as causing impacts to the project, may have longer-term effects, as negative feelings will usually be buried by those who feel them. This happens all the time in both business and personal life. Unless this unresolved conflict is resolved at some point, it will continue to fester and even build, until at some point it becomes too big for the person to keep buried and an explosion is likely to occur.

There are a number of methods that are recommended for dealing with conflict, and each of these is quite appropriate for use under some conditions. Many different professionals have proposed various methods and ways of dealing with conflict. Some of these suggest categories that are similar to those suggested by

Thomas and Kilman, who describe five techniques. Only one of these, however, can fully resolve the underlying conflict behind the situation that has caused the problem. That one, of course, is problem solving, since this is the only one that addresses and resolves the underlying issues. An important lesson here is that there is not just one solution. Different methods are sometimes appropriate. Conflict in a military hierarchy, for example, is not typically resolved by collaboration.

Wikipedia, which is consistent with other reference material, has a good summary of the Thomas and Kilman techniques:*

Five basic ways of addressing conflict were identified by Thomas and Kilman in 1976.

- *Accommodation*—surrender one's own needs and wishes to accommodate the other party.
- *Avoidance*—avoid or postpone conflict by ignoring it, changing the subject, etc. Avoidance can be useful as a temporary measure to buy time or as an expedient means of dealing with very minor, non-recurring conflicts. In more severe cases, conflict avoidance can involve severing a relationship or leaving a group.
- *Collaboration*—work together to find a mutually beneficial solution. While the Thomas Kilman grid views collaboration as the only win–win solution to conflict, collaboration can also be time-intensive and inappropriate when there is not enough trust, respect or communication among participants for collaboration to occur.
- *Compromise*—find a middle ground in which each party is partially satisfied.
- *Competition*—assert one's viewpoint at the potential expense of another. It can be useful when achieving one's objectives outweighs one's concern for the relationship.

The Thomas–Kilman Instrument can be used to assess one's dominant style for addressing conflict.

LEARNING ORGANIZATION

Learning organization is a term coined by Peter Senge. The concept of a learning organization is relatively new in project management and probably more so in telco environments. However, this concept is completely consistent with project management, since one theme of project management is learning from previous mistakes or problems. A learning organization is one that encourages people to learn from past mistakes. In order for this to happen, the organization has to allow mistakes to occur and ensure that people share information about the problem and the solution, without recrimination for things that go wrong. Project management can thrive much better in learning organizations.

*http://wikipedia.org/wiki/conflict#cite_note-6. Accessed June 2010.

MANAGING THE WORKLOADS OF THE TEAM MEMBERS

While we are talking about people, we need to consider the work that each has been assigned. As we have seen, planning the project is an iterative process with many variables to consider. Even the initial development of the schedule is quite complex, and it can be problematic even before resource assignments are fully considered. In the initial planning phase, the focus is on the project scope, objectives, timing considerations, and cost. Later, the loading of the team and the team members is usually taken into consideration, and the schedule or the resource assignments may be rearranged in order to optimize both. Generally, the project work is reasonable from a resource viewpoint at the beginning at least. But then, of course, work starts, and inconvenient reality sets in. Schedules become shortened. Problems cause work to take longer than planned, creating a time crunch for subsequent activities, and changing the initial times at which different people will be needed on specific tasks. When this happens, the PM really needs to do another assessment of the loading over time of each of the resources. Has someone been overloaded at any point in time, especially in cases in which the people work on more that just the project work? What will be the impact of that level of loading on that person? Can anything be done to alleviate any problems? How can this best be done with the team, to ensure that everyone continues to support the project and be happy about the situation? The PM needs to spend time considering this. The team members will appreciate this concern for their welfare, and be more likely to step up to the plate. I have even seen staunch union members willingly take on extra work without requesting extra compensation for it, to help the team to meet their goals, because they knew that their work and contributions were appreciated, and they felt like members of the team.

SOME SUGGESTIONS FOR GOOD PARTICIPATION AND GOOD MANAGEMENT

A good project manager might not have background in all of the topics discussed briefly in this chapter, but the more knowledge and skill a manager can have in these areas, the better he or she is likely to be as an effective project manager. In addition to understanding the concepts, though, the right behavior and follow-through is needed for effective management. Here are a few tips, collected from numerous sources including first-hand experience, that will help a PM to be effective:

1. Build and manage relationships with the team members and the stakeholders.
2. Be honest, sincere and trustworthy. People will not follow someone they do not trust.
3. Communicate clearly, effectively, and frequently so that there are no misunderstandings. Good communications techniques were discussed in Chapter 10, but a few suggestions also follow here:

- Develop and use good listening skills.
- Be professional and objective, keeping emotions out of business communications unless there is some reason that they need to be included.
- Keep communications clear and assertive so that people understand what you need and that you are serious in looking for it.
- Keep a positive focus on the communications.
- Ensure that you get feedback; follow up and use redundancy to ensure that the right message gets across.

4. Care about the team members and the stakeholders. They will be able to tell that you care and this will make a big difference in the way they will do the work. Empathize with team members. Define the scope clearly, as well as the roles and responsibilities, to minimize the potential for misunderstanding.

CHAPTER 15

WHAT IS THE GAIN?

Up to this point, we have looked at the telecom environment from many different perspectives. We should have a good understanding of the issues companies and individuals are faced with in this environment. We have also looked at the basic areas of project management, and should have some insight into how these apply in general as well as why they are needed in a telecom environment. This environment today is one in which there is a lot of pressure and significant sense of urgency in all aspects of the work. The application of project management has a cost in time with a corresponding financial cost to pay for time that is used for work that is not directly productive. Given this, we now need to ask whether or not, in this environment, it is worth the effort, time, and cost to apply project management. This text presents a wide range of project management techniques. In many cases, a PM may decide that some of the techniques discussed are not worth the cost to use them on a specific project. This decision must be based on the full picture. Hopefully, the material in this book has given the reader enough insight to allow for good decisions regarding the use of the different techniques, with enough illustration of the benefits of using the techniques and the pitfalls of not using them. The PM can then weigh these against the costs to make a well-informed decision.

GOOD PEOPLE

In summary, let us enumerate the things that are most important for a PM to consider, and think about what needs to be in place to improve the success potential for a particular project. One of the biggest factors in project success is the people who work on the team. The selection of the right team can be one of the biggest factors for success. The PM must select the right good people. Not only must these people have the right technical (read: functional) abilities for every assigned project activity, but they must be people who are motivated to make the project a success. They must be people who have the ability and the motivation to understand not only the project and its goals, but also how it fits into the whole current environment and corporate direction. They must be people who can work well with others, specifically the others who will be on this particular project team. They must also be responsible people who will step up to all the requirements of the project. If the PM has the opportunity to select the team members, and does this wisely, the project will be better positioned for success.

UNDERSTANDING THE VALUE OF THE PRODUCT

The sponsor, the PM, and the team must also have a very clear understanding of the required results from the project. They need to understand how the project and the product that it produces will impact the sponsoring company and any external stakeholders. The full team needs this understanding to enable each member to think productively and creatively about actions, directions, and solutions that will enhance the output of the project. With an understanding of the true value the project brings, the team can become more committed and can work smarter toward the right objectives and goals.

CLEAR PROJECT OBJECTIVES

The project must have clear objectives, which the sponsor, the PM, and the team all understand, and there must be consistent understanding of these amongst all the project stakeholders. To achieve this, communication and focus on the objectives is needed. This cannot just happen on its own. If team members do not understand what the objective is, they cannot meet it; if stakeholders understand the objective differently than the team members, meeting the objective may not provide success.

CLEARLY DEFINED SCOPE

Much of this book has addressed ensuring that there is a clear definition of the project scope. Project management techniques provide many views of the scope, from the charter to the scope description document to the work breakdown structure. Unless there is a very clear understanding of the desired scope of the project, in all areas including quality expectations, the team will have great difficulty in providing the right output. Time will be lost if team members find themselves debating what is it that they should be producing. Conflict in understandings is quite likely to erupt if the definition is not clear, so a clear scope description is definitely one of the first requirements for project success. A clear scope definition requires the project to have clearly defined deliverables. The work breakdown structure gives the first pass at this, but the scope description should also provide details explaining each required deliverable.

GOOD PLANNING

Thorough and accurate planning is very important in ensuring that projects will be successful. This includes scope planning as mentioned above, but good planning must cover other process areas as well. The project requires solid financial planning, to create a realistic, workable budget and then control it. The schedule needs to be well planned, monitored, and controlled, with clear links between schedule, scope, and expense. In addition to clear communication of the objectives, it is very important that the team members plan and then implement regular, clear, and complete communication about all aspects of the project, to all project stakeholders. We have seen that planning for the project risks, ways to deal with these risks, and the creation of appropriate contingencies are also necessary for project success. Planners must analyze the particular situation of a given project and derive an overall master plan appropriate for that project, taking into consideration:

- Project objectives
- Risk and uncertainty management
- Deliverables and activities
- Project cost
- Determination of resource requirements
- Communication amongst the team members and the stakeholders
- Possible alternative courses of action
- Measurement and reporting of genuine progress
- Basis for making decisions and any required corrective action

STRONG CHANGE CONTROL

Another really critical input to project success is strong change control. The team needs to ensure that there is a process in place to define all aspects of control for proposals for change, so that changes, whether originating from new ideas put forward by the stakeholders, "feature-creep," late deliverables, budget cuts, and similar hazards are not implemented blindly. A telecom project is generally a complex undertaking, and changes in one area will have impacts in others. A formalized process of evaluating a proposed change is very useful in ensuring that the impacts of the change are clearly understood by all stakeholders, and that a well-considered decision can be made. Most significant changes will drive a need for additional time, funding, or resources, and ensuring that these needs can be met is part of the decision process. Many proposed changes are ill-considered, with impacts in other areas not understood. It is the PM's job to make sure that all the necessary information is available for a correct decision. Sometimes, the Emperor really does not have any clothes, and it falls to the PM to point this out.

WELL-CONNECTED TEAM

Teamwork is another basic ingredient for success. A group of people with the right skills could well produce a product, and perhaps do this within all the project constraints. But with effective teamwork, this can be achieved much more easily, more quickly, and will provide higher quality results.

EFFECTIVE COMMUNICATION

Underpinning it all is the need for clear effective communication. Constant and effective communication, including management of the expectations of the client and other key stakeholders must occur throughout the run of the project.

With especially the items mentioned in this chapter, and certainly with the implementation of all the techniques and processes discussed in this book, the team can maximize the magnitude and the chances of success for projects.

Thinking back through the nature of the environment, the speed of the changes, and the urgency of the projects, it should be clear that each of the project management techniques and tools can help a team in any of the companies in the new electronic communications environment to be more successful in delivering products and services that are better, more timely, and within budget, while at the same time keeping stakeholders happy. That is what project management is all about.

BIBLIOGRAPHY

A Guide to the Project Management Body of Knowledge, PMBOK® Guide, Fourth Edition, Project Management Institute, 2008.

Kerzner, H., *Project Management: A Systems Approach to Planning, Scheduling and Controlling,* 9th ed., Wiley, 2005.

Kerzner, H., *Project Management: A Systems Approach to Planning, Scheduling and Controlling,* 10th ed., Wiley, 2009.

Desmond, C., *Project Management for Telecommunications Managers,* Kluwer Academic Publishers, 2004.

J. R. Meredith and S. J. Mantel Jr., *Project Management, A Managerial Approach,* Seventh Edition, Wiley, 2009.

INDEX